Sustainable Island Tourism

Competitiveness and Quality of Life

CABI Series in Tourism Management Research

General Editors:
Professor Eric Laws, Professor of Tourism, Siam University, Bangkok
Professor Noel Scott, Griffith Institute for Tourism, Griffith University, Australia

Since the mid-20th century, modern tourism has grown rapidly in extent and diversity, becoming increasingly competitive and volatile as it is impacted by climate change, new technologies, changing distribution systems and the opening of new markets. As a result, governments, tourism destinations and businesses need to improve their management capability and adopt best practices to survive. The purpose of this series is to provide tourism managers, administrators, specialists and advanced students with state-of-the-art research and strategic knowledge to enable them to thrive in dynamic and unpredictable environments. Contributions are based on critical and interdisciplinary research that combines relevant theory and practice, while placing case studies from specific destinations into an international context. The series presents research on the development and diffusion of best practice in business and destination management that fulfils the objective of environmental, sociocultural and economic sustainability at both the local and global scale.

The cover design for this series depicts a generalized mosaic composed of many tiles. Metaphorically, this illustrates our philosophies that while the various elements of tourism require specific study, it is the overall picture that is most significant, and that tourism is a very dynamic, complex and evolving industry. This series seeks to build a coherent approach to future tourism research through each individual title.

Titles available

1. *Tourism Crisis and Disaster Management in the Asia-Pacific*
Edited by Brent W. Ritchie and Kom Campiranon

2. *Destination Competitiveness, the Environment and Sustainability: Challenges and Cases*
Edited by Andres Artal-Tur and Metin Kozak

3. *Visitor Management in Tourism Destinations*
Edited by Julia N. Albrecht

4. *Sustainable Island Tourism: Competitiveness and Quality of Life*
Edited by Patrizia Modica and Muzafer Uysal

Sustainable Island Tourism

Competitiveness and Quality of Life

Edited by

Patrizia Modica

Department of Economics and Business, University of Cagliari, Italy

and

Muzaffer Uysal

Department of Hospitality & Tourism Management, University of Massachusetts, USA

CABI

CABI is a trading name of CAB International

CABI
Nosworthy Way
Wallingford
Oxfordshire OX10 8DE
UK

Tel: +44 (0)1491 832111
Fax: +44 (0)1491 833508
E-mail: info@cabi.org
Website: www.cabi.org

CABI
745 Atlantic Avenue
8th Floor
Boston, MA 02111
USA

T: +1 (617)682-9015
E-mail: cabi-nao@cabi.org

A catalogue record for this book is available from the British Library, London, UK.

Library of Congress Cataloging-in-Publication Data

Names: Modica, Patrizia, editor. | Uysal, Muzaffer, editor.
Title: Sustainable island tourism : competitiveness, and quality-of-life / edited by Patrizia Modica, Department of Economics and Business, University of Cagliari, Italy and Muzaffer Uysal, Pamplin College of Business, Virginia Polytechnic Institute & State University, USA.
Description: Wallingford, Oxfordshire, UK ; Boston, MA, USA : CAB International, [2017] | Series: CABI series in tourism management research | Includes bibliographical references and index.
Identifiers: LCCN 2016031183 (print) | LCCN 2016050000 (ebook) | ISBN 9781780645421 (hbk : alk. paper) | ISBN 9781780645438 (ePDF) | ISBN 9781786391308 (ePub)
Subjects: LCSH: Sustainable tourism. | Islands--Environmental aspects.
Classification: LCC G156.5.S87 S64 2017 (print) | LCC G156.5.S87 (ebook) | DDC 910.68/4--dc23
LC record available at https://lccn.loc.gov/2016031183

ISBN-13: 978 1 78064 542 1

Commissioning editor: Claire Parfitt
Editorial assistant: Emma McCann
Production editor: Shankari Wilford

Typeset by AMA DataSet, Preston, UK
Printed and bound in the UK by CPI Group (UK) Ltd, Croydon, CR0 4YY, UK

Contents

Editors and Contributors

Editors

Patrizia Modica is Associate Professor in Business Economics at the University of Cagliari, Italy, where she also teaches issues relating to the tourism industry. Within the field of tourism, her interests are focused on tourism sustainability. Author of a number of publications, including the book, in 2015, *Sustainable Tourism Management and Monitoring. Destination, Business and Stakeholder Perspectives*, she has presented at conferences related to the managerial aspects of tourism, including the 2014 OECD-EUROSTAT Conference in Nara, Japan. In her role as Destination Coordinator of the DMO project Visit South Sardinia, within the ETIS (European Tourism Indicator System) programme, she has been awarded by the European Commission with the 2016 European Sustainable Destination prize of the first joint ETIS and Accessible Tourism Awards. E-mail: modica@unica.it

Muzaffer Uysal is a professor and chair in the Department of Hospitality and Tourism Management at Isenberg School of Management at University of Massachusetts. He is a member of the International Academy for the Study of Tourism, the Academy of Leisure Sciences, and serves as co-founder of *Tourism Analysis: An Interdisciplinary Journal*. In addition, he sits on the editorial boards of more than ten journals, including *Journal of Travel Research* and *Annals of Tourism Research*. He has authored and co-authored numerous articles, monographs, and several books related to tourism research methods, tourist service satisfaction, tourism and quality-of-life, experience value in tourism, tourism-related scales, and management science applications in tourism and hospitality. Dr Uysal has received a number of awards for research, excellence in international education, teaching excellence, and best paper awards. His current research interests focus on tourism demand/supply interaction, tourism development, and quality-of-life research in tourism. E-mail: muysal@isenberg.umass.edu

Contributors

Levent Altinay is a Professor of Strategy and Entrepreneurship at the Oxford School of Hospitality Management at Oxford Brookes University. Levent's research interests are in the areas of entrepreneurship and strategy. Using primarily qualitative methods as well as mixed methods, he is particularly interested in how entrepreneurs start up and develop their businesses and how firms establish partnerships internationally. Levent sits on the editorial boards of more than 12 journals including the *Journal of Business Research* and *Journal of Services Marketing*. He is the Associate Editor, Europe, for the *International Journal of Contemporary Hospitality Management*. Levent is also the co-author of hospitality management research methods, entrepreneurship and strategic management text books: *Planning Research in Hospitality and Tourism* (2015); *Entrepreneurship in Hospitality and Tourism* (2015); and *Strategic Management for Hospitality and Tourism* (2010). E-mail: laltinay@brookes.ac.uk

Meghan Beardsley serves as a research associate in the Department of Hospitality and Tourism Management in the Pamplin College of Business at Virginia Tech. She is currently pursuing her PhD from the Department of Hospitality and Tourism Management. Her primary areas of research interest include destinations in transition, tourism development, government and international affairs, and events evaluation and education. E-mail: cinemeg2@vt.edu

Kelly S. Bricker is Professor and Director of Parks, Recreation, and Tourism in the College of Health at the University of Utah, USA. Dr Bricker conducts research related to the impacts of tourism on protected areas and communities. She also studies sense of place and the contribution of parks and protected areas as health resources. She has published and has been invited to present internationally on issues in ecotourism certification and policy, sense of place, natural resource management issues, and the relationships between health and sustainable tourism. Recently, with Rosemary Black and Stuart Cottrell, she authored and edited a book titled *Sustainable Tourism and the Millennium Development Goals: Effecting Positive Change*, and *Demystifying Theories in Tourism Research*, with Holly Donohoe. E-mail: kelly.bricker@health.utah.edu

Jack Carlsen is Founder and Director of Tourism Research Services, providing tourism and economic development research, planning and management services in Australia and internationally since 1991. He is also an Adjunct Professor at Curtin University (Perth), where he established the Curtin Sustainable Tourism Centre in 2002 in his role as Professor of Sustainable Tourism. He has achieved more than 150 academic and technical publications related to tourism, events and sustainability research, planning and management. E-mail: jackcarlsen2@bigpond.com

Robertico Croes currently serves as the Associate Dean and the Associate Director of the Dick Pope Sr. Institute for Tourism Studies at the Rosen College of Hospitality Management at the University of Central Florida. Professor Croes is the author of three books pertaining to tourism demand and competitiveness in small island destinations as well as tourism development and poverty alleviation. He has published more than 100 journal articles, book chapters and conference presentations in the fields of tourism demand analysis/forecasting, tourism economic impact, competitive and sustainable tourism in tourism development analysis, tourism development applied to poverty alleviation and quality of life in small islands and developing countries. E-mail: robertico.croes@ucf.edu

Anna Farmaki is a Lecturer in Tourism Management at the Department of Hotel and Tourism Management at the Cyprus University of Technology. Her research interests lie in the areas of tourism planning and development and tourist behaviour. E-mail: anna.farmaki@cut.ac.cy

Mark P. Hampton is Reader in Tourism Management at the University of Kent, specializing in small-scale/coastal/island tourism in South-east Asia. He publishes regularly in leading journals and his books include *Backpacker Tourism and Economic Development*; and *Tourism and Inclusive Growth in Small Island Developing States* with Julia Jeyacheya. He has extensive field experience (South-east Asia, Caribbean, Indian Ocean) funded by: the World Bank; Commonwealth Secretariat; Foreign and Commonwealth Office; DFID; Swiss overseas aid; Malaysian Ministry of Tourism; and the British Academy. He is a Fellow of the Royal Geographical Society and Visiting Professor at Universiti Teknologi Malaysia. E-mail: m.hampton@kent.ac.uk

Julia Jeyacheya is Lecturer in Project Management at Bradford Centre for International Development, University of Bradford. She has co-authored several policy-led reports for international organizations including the World Bank, and a book for the Commonwealth Secretariat (2013) with Mark Hampton, has published articles in leading international journals such as *World Development and Review of International Political Economy*, and has extensive fieldwork experience in Botswana, Namibia, Zimbabwe, the Seychelles, Vietnam and Myanmar. She is a Fellow of the Royal Geographical Society. E-mail: j.jeyacheya@bradford.ac.uk

Deborah L. Kerstetter is Professor of Recreation, Park and Tourism Management in the College of Health and Human Development at the Pennsylvania State University, USA. Dr Kerstetter conducts research related to the underlying factors affecting individuals' travel decisions. She also studies the cultural and social impacts of travel and tourism on individuals and communities. Her nearly 100 manuscripts have been published in highly respected journals such as the *Annals of Tourism Research*, *Tourism Management*, the *Journal of Travel Research*, the *Journal of Sustainable Tourism* and *Tourism Analysis*. She has written multiple chapters for books devoted to the study of tourism and has been invited to present the results of her research at conferences around the world. E-mail: debk@psu.edu

Marketa Kubickova is an Assistant Professor at the University of South Carolina in the School of Hotel, Restaurant and Tourism Management. Dr Kubickova obtained her Doctoral degree from the University of Central Florida. Prior to joining academia, she held a number of managerial positions in several upscale international hotels. Her research interests mainly focus on competitiveness, hotel sustainability and the role of government, with special emphasis on developing countries. Her research has been published in numerous tourism and hospitality journals. She is a member of American Hotel and Lodging Association, Women in Lodging, and Delta Epsilon Iota. E-mail: kubickova@hrsm.sc.edu

Seungwoo John Lee is an Associate Professor at Sangmyung University in the Department of Foodservice Management and Nutrition, South Korea. His research interests include volunteer tourism, sustainable tourism, multi-stakeholders' perception of tourism development and food service management. E-mail: slee@smu.ac.kr

Haywantee Rumi Ramkissoon is Associate Professor and Director of Tourism Research Cluster at Curtin Business School, Perth. She is also a senior research fellow at Monash University. Rumi holds two doctoral degrees in Tourism and Applied Environmental Psychology. Her post-doctoral experience relates to place attachment, and societal innovation and behaviour change. Rumi publishes in leading tourism journals such as *Annals of Tourism Research*, *Tourism Management*, *Journal of Travel Research*, *Journal of Sustainable Tourism*, *Tourism Analysis* and *Journal of Hospitality and Tourism Research*. She is the book review editor for *Current Issues in Tourism* and Research Note Editor for *Journal of Hospitality Marketing and Management*. Rumi serves on 13 editorial boards of high quality journals in her field. E-mail: haywantee.ramkissoon@curtin.edu.au

Barbara Terenzi is an anthropologist and coordinator of the Comitato per la promozione e protezione dei diritti umani and the Human Rights and Advocacy Department of VIS (International Volunteers for Development). She specialized on Nepali shamans in the Limbu Region, besides participating in numerous EU and UN development projects around the world, and has coordinated various field researches on cultural heritage focused on religious events in southern Italy. Since 2007, she has been mapping Sardinia cultural heritage with special focus on the ritual of perpetual vow fulfilment and the feast of Sant'Efisio, elaborating the dossier for inclusion in the UNESCO Representative List of the Intangible Cultural Heritage of Humanity. She is the co-editor of many publications, among which are: *Arcipelago Mediterraneo: la Sardegna and Feste e riti d'Italia, Sud 1*; *Goals for Mankind*, a report to the Club of Rome. E-mail: barbara.terenzi@gmail.com

Allan Wright is a Senior Economist at the Central Bank of Barbados and also currently a Researcher at the Center for Monetary Studies in Latin America (CEMLA). He received his PhD in Economics from the University of the West Indies. His current work lies mainly in the areas of debt and fiscal policy, forecasting, economic growth, foreign direct investment and aspects of the real sector including tourism. E-mail: wright@cemla.org

Muhammet Yaşarata has a PhD from North London University and is currently the owner of Cyprus Premier and Cyprus Paradise tour operation businesses. He is also the President of Cyprus Turkish Chamber of Commerce in the UK. He is interested in research on sustainable tourism, politics, development and planning of tourism in small island economies. E-mail: muhammet@cyprusparadise.com

Mariyam Zulfa completed her PhD at Curtin Business School, Curtin University, Perth, in 2009. Her thesis focused on competitiveness of small islands tourism destinations. She served as Maldives' Minister of Tourism, Arts and Culture from 2010 to 2012. As of 2013, she was admitted to the Supreme Court of Western Australia as a barrister and solicitor and is currently working in private practice and reading international law (part-time). E-mail: mariyamzulfa@gmail.com

Foreword

Sustainable tourism encompasses a broad set of techniques and advanced methodologies for analysis and policy making with regard to sustainability, competitiveness, quality of life and satisfaction.

Sustainable Island Tourism: Competitiveness and Quality of Life presents an interdisciplinary integrated approach, with a rich and stimulating set of case studies and specific chapters, with a special focus on small islands.

A heartfelt thanks to Patrizia Modica and Muzaffer Uysal for having involved me in this endeavour; it is an honour to have had the opportunity of bringing a small contribution to this work, which, I believe, will surely enrich the existing literature on these matters.

The perception of distance, nowadays, has substantially changed both in terms of space and time. In a globalized and accelerated world, news and images travel around with almost no limitation, often in the same instance when an event takes place, and have an impact also on the use of leisure time and the quest for encountering new and different environments.

Travelling, therefore, in time has ended up by assuming new features and even stronger influence on people's life. In addition, the way tourism often facilitates a face-to-face encounter between populations with different cultural backgrounds can be evidenced, henceforth, opening a window on differences – an important reflection currently in progress while considering the flow of migrants who are bearers of their specific cultural heritage, with traditions and savoir faires that somehow impact also on the local communities hosting them.

Islands, in such discourse, offer the opportunity for an interesting living adventure, in particular those with specific characteristics such as dimensions and cultural and morphological aspects that have preserved their unique features.

'Insularity is not only an environmental condition but a social situation, a potential symbol in a cultural geography' highlights Robb (2001), to which Knapp (2007) adds 'a multi-faceted, dynamic and changing concept not only for those whose everyday life is conditioned by it, but also for those who employ it as an intellectual construct'.

Often small islands, because of their contained aspect, confined within their water boundaries, maintain a socio-economic and cultural breadth which has somehow safeguarded aspects of everyday life, exercising a strong attraction and fascination. As also UNESCO (2003) has evidenced: 'Small islands as cultural crossroads. The history of the islands . . . testifies to the great cultural interaction and mixing that small islands have offered humanity'.

Time, rhythm of life, relations, nature, culture, all features of a different way of living, which are often opposed to the quality of life faced in the huge metropolis of mainland and the alienation to it attached as repeatedly emerges from the work of MacCannell (1976).

In addition, the same action of leaving our protected and well-known milieu to enter into a new dimension, different in geographical but also cultural and socio-economic terms, is by itself a journey; a journey within ourselves that can lead to discovering insights and resources.

It can be worth observing such process along a set of main passages constituting different aspects of the endeavour undertaken.

The first phase will regard the initial step of leaving a reality immediately known, where no mediation is necessary to understand the surroundings and its connected codes. It represents the action of stepping out of the daily routine to engage into a journey, physical and mental, leading into a dimension that is different, where exploration and flavouring diverse emotions is the challenge through which we lose ourselves and return, maybe, modified.

An intermediate passage follows: the physical journey from one place to the other. This can be considered as a liminal dimension, when time is suspended waiting for arrival and the elaboration is an imaginary exercise on building up expectations. In addition, burdens are left behind and a feeling of freedom emerges in preparing for encountering an 'other' dimension with different rules, constraints, codes together with the curiosity and excitement for the unknown and the surprise of discovery.

Finally, once the destination is reached, a new stage is initiated. Here encounter and exploration occur, an opportunity to discover new internal resources and capacities and at the same time a challenge of living out of the boundaries of a well-known milieu. Certainties have been left behind in favour of new emotions and environments; an experience of enrichment that can lead towards new perspectives or open a different view about life: a treasure in experience, a richness to bring back.

The journey, in these terms, can be then cathartic. It may be compared to the process that takes place during a pilgrimage: a preparation to step out, a liminal passage, the experience of the sacred and the return to normal life modified by the exposure (Terenzi, 2009).

The search for knowledge possibly will become the force triggering the capacity of savouring strong emotions which normal everyday routine has quelled. An occasion for discovering the richness of a cultural heritage that is different, hence facilitating the understanding of the potentiality inherent in diversity.

In such perspective, tourism plays a pivotal role, sustainable tourism in particular with its attention to helping destinations in safeguarding their uniqueness and, at the same time, the capability of putting together those who arrive with those who host, allowing both to share an understanding, which might be of growth, and build up exchanges with mutual advantages.

Small islands, therefore, offer a significant opportunity and a wealth as, quoting again UNESCO:

> . . . is reflected in the priority given in certain regions to the preserving of traditions, valuing local languages and promoting the identity and dignity of islanders. The incorporation of the knowledge and wisdom of local users and interest groups in the formulation and sharing of wise management practices is one way to strengthen these dimensions

On the other hand, as Ismail and Turner (2008) point out, 'the influx . . . of tourists to an island destination is likely to have a profound effect on the destination in cultural, social and environmental terms because of the destination's small size'; a reflection also reiterated by Colin and Baum (1995) while considering the peculiar value of small dimension islands: 'the local population is more likely to have frequent contact with tourists compared with larger islands or mainland destinations'.

Finally it is worth here recalling that sustainable tourism and the role played nowadays by small islands are expressly mentioned elements of the recent United Nations 2030 Agenda for Sustainable Development with its Sustainable Development Goals (SDGs)(United Nations, 2015). In particular SDG 8.9 invites to 'devise and implement policies to promote sustainable tourism that creates jobs and promotes local culture', while the attention to small islands is mainstreamed in most of the SDGs.

At last, from immemorial time, man has always travelled in search of more intense emotions and to discover new and diverse horizons, but this, also is in pursuit of the return, as Ulysses teaches.

Barbara Terenzi, anthropologist
Rome, January 2016

References

Colin, M. and Baum, T. (eds) (1995) *Island Tourism. Management Principles and Practice.* Wiley, Chichester, UK, pp. 33–51.

Ismail, F. and Turner, L. (2008) Host and tourist perceptions on small island tourism: a case study of Perhentian and Redang Islands, Malaysia. In: Tsounis, N., Vlahvei, A., Monovasilis, T. and Siskos, E. (eds) *Proceedings: International Conference on Applied Economics* – ICOAE, Kastoria, pp. 401–410.

Knapp, A.B. (2007) Insularity and island identity in the prehistoric Mediterranean. In: Antoniadou, S. and Pace, A. (eds) *Mediterranean Crossroads.* Pierides Foundation, Athens, pp. 37–52.

MacCannell, D. (1976) *The Tourist*, 2nd edn. Stocken, New York.

Robb, J.E. (2001) Island identities: ritual, travel and creation of difference in Neolithic Malta. *European Journal of Archaeology* 4, 175–202.

Terenzi, B. (2009) Culture viventi. In: Massari, S. (ed.) *Feste e Riti d'Italia. Sud 1.* De Luca Editore, Isituto Centrale per la Demoetnoantropologia, MiBACT, Rome, pp. 46–57.

UNESCO (2003) Intersectoral Platform on Small Islands Developing State, Small Islands: Looking Forward – Beyond 2004. 6. Promoting island identity to effect social change. Available at: http://portal.unesco.org/en/ev.php-URL_ID=13945&URL_DO=DO_TOPIC&URL_SECTION=201.html (accessed 13 January 2016).

United Nations (2015) *Transforming our world: the 2030 Agenda for Sustainable Development,* Resolution adopted by the General Assembly, 25 September 2015, A/RES/70/1, UN, New York.

Preface

Over the past 40 years the field of tourism has witnessed a tremendous growth in the number of academic journals, books and the amount of information that has been generated on different aspects of tourism, including sustainability in general. As tourism displays maturity and scientific sophistication, it is important that we as tourism researchers fully understand and further examine the breadth and depth of the salient aspects of island tourism issues such as sustainability, competiveness and quality of life of stakeholders that engage in production of tourism goods and services and planning and development decisions. Although there has been a significant number of books on sustainability in tourism, there is a limited number of books that have specifically focused on sustainable island tourism issues. Of these books, none has directly addressed the interplays of sustainability, competiveness and quality of life as important constructs of tourism development with respect to island destinations. In this regard, we believe that this edited volume is an important contribution in filling this particular gap. The aim of this book is to provide a portfolio of conceptual, empirical and case-based studies to address the interplays of sustainability, competitiveness and quality of life in islands destinations. We have brought together 18 scholars representing diverse areas to address the nature and types of sustainable island tourism and how the elements of competiveness and quality of life of stakeholders are intimately connected to understand and examine challenges and the long term success of tourism development and its business activities.

The main themes and aspects are investigated through a case studies methodology focused on a set of significant islands in the world, analysing their specific features with the aim of putting together examples with different backgrounds and geographical locations: Jeju Island, South Korea in the Korean strait of the East China Sea, the Perhentian Islands, Malaysia in the Siam Gulf, the Maldives in the Indian Ocean, Cyprus and Sant'Antioco (Sardinia) in the Mediterranean Sea, Cuba where the Caribbean Sea, the Gulf of Mexico and the Atlantic Ocean meet, and the small island developing states within the Caribbean and Fiji in the Pacific Ocean.

The publication highlights the characteristics of island tourism development, specifically outlining different impacts on the natural environment and local economy, including its effect on employment, quality of life, culture and traditions, with a set of strategic indicators suggested to plan, monitor and control the sustainability of island tourism destinations. The book also explores different public and private instruments for supporting the sustainable tourism development in island destinations. Finally it explores different strategies for improving competitiveness of island destinations and increasing advantages for all stakeholders. For these reasons the book would be of great interest to the students of tourism, practitioners, tourism planners and developers, decision makers and the like. In addition, researchers of tourism and recreation would find this endeavour very useful in understanding how best to cater to, attract and increase tourists, since it focuses on the merits and importance of the interplays of sustainability, competitiveness and quality of life of island stakeholders. Enjoy it!

Patrizia Modica
Muzaffer Uysal

Part I Sustainable Island Tourism

1 Sustainability and Tourism Development in Island Territories

Patrizia Modica[1]* and Muzaffer Uysal[2]

[1]*University of Cagliari, Sardinia, Italy;* [2]*University of Massachusetts, Amherst, USA*

1.1 Introduction

This chapter analyses the concept of sustainability and island development, related problems and constraints in tension between heritage and landscape conservation and economic development. What type of tourism in island destinations is more suitable? Either mass or elite tourism, both need a solid base of sustainable principles and management practices. Islands are very peculiar, due to several circumstances, e.g. cultural and political. Moreover, the condition of isolation makes these destinations unique places in the world, with regard to the natural environment and related ecosystem, which deserves attention to guarantee a balance between the need for both conservation and development. Specifically, tourism development has been considered a viable green economic growth for decades, since the WBGU (1996) declared mass tourism as one of our planet's seven syndromes. The unique and fragile *equilibrium* in the natural, social, cultural and economic domains that characterizes island destinations devoted or developing tourism activities, constitutes the reason to undertake the current study.

What does *equilibrium* for *island destinations* mean and how can it be measured? The notion of equilibrium has been referred, in literature, to complex systems in tourism (Faulkner and Russell, 2001; Baggio, 2008; Yang and Sun, 2013) and destination networks (Scott *et al.*, 2008; Beaumont and Dredge, 2010) but, with reference to island destinations, it is relatively little studied (Pechlaner *et al.*, 2003). Considering the three pillars in which sustainable tourism is usually regarded (UNWTO, 2004; UNEP and UNWTO, 2005; ILO, 2010), it seems that the environmental, social and economic aspects are all essential to discuss about equilibriums in island destinations. These aspects of sustainability are substantially meaningless if the managerial aspect in sustainable tourism (Miller and Twining-Ward, 2005) is not addressed (Modica, 2015). The branch of literature addressing sustainability in complex systems, such as

*E-mail: modica@unica.it

destinations, has been developed in the last decades in terms of adaptive management (Schianetz *et al.*, 2007; Plummer and Fennell, 2009; Karatzoglou and Spilanis, 2010; McCook *et al.*, 2010; Larson and Poudyal, 2012). In general, a place shows its natural, historical, cultural and basic elements to attract visitors, none the less the management of tourism in destinations can have positive or negative effects upon the environment to be preserved, or developed, in the perspective of future existence and prosperity in the medium to long term. In this regard, McMahon (1999) affirms 'no place will retain its special appeal by accident'. Tourism can be either the engine of floridity or degradation, depending on the way in which it is managed at the different levels, supra-national, national or sub-national (UNWTO, 2004; Hall *et al.*, 2015).

The concept of equilibrium, and its opposing situation – disequilibrium – in island destinations can be expressed considering two prevalent tourism theories: the Carrying Capacity (K) and the related theorization of the Limits of Acceptable Change (LAC) for a punctual framework on the balance of the previous mentioned opposing situations (Ahn *et al.*, 2002; Lusseau, 2008) and the Tourism Area Life Cycle (TALC) theorized in 1980 by Butler. TALC in this chapter becomes functional to relate the key variable of destination equilibrium to the specific phases that tourism areas can pass through from the initial stage of exploration, when visitors start to explore an area, to the possible final stage of decline, when a resort area has constantly been abandoned by tourists and tourism services (Hall, 2006; Lagiewski, 2006).

1.2 Theoretical Approach of the Study

Tourism areas as islands encounter major or minor difficulties in the development of tourism activities, compared to mainland territories, depending on the aspects of sustainability to be managed. The geographical and morphological peculiarity of a land facing the sea all over its boundaries can be both a strength for the preservation of the natural and cultural heritage and a weakness for island economic development due to the restrictions in the attribute of destination accessibility. Accessibility in a broad sense can be defined as the possibility for a population to access a destination 'via road, air passenger services, rail or cruise ships' (UNWTO, 2007, p. 2). In general, the more or less 50,000 islands in the world are not accessible via road or railway; this allows us to regard the island as a world on its own, either sustainable for impediments to reach it – and the consequent low human pressure on the environment – or not sustainable due to the objective limitation in its accessibility. The preservation of a destination as an island and the natural development of visitor services in the different stages of a TALC highlight the compromise between two opposite situations and their management in the perspective of equilibrium of territories, people and organizations. The theory of Carrying Capacity appears suitable to investigate the application of the theme equilibrium (Holden, 2008; Lusseau, 2008) in islands as tourism destinations (Kersetter and Bricker, 2012). Another theory in literature seems strictly related to the discussion on island tourism and its relationship with the key aspects of sustainability, competitiveness and quality of life developed in this chapter, the Limit of Acceptable Change theorization (Frauman and Banks, 2011), considered in literature as the operationalization (Ahn *et al.*, 2002) of the more speculative theory of the Carrying Capacity. The TALC has the potential in this book to contextualize islands, which are analysed through case studies on prevalently small island destinations, depending on the phase of their life cycle of development. With TALC, Butler (1980) develops a pattern based upon the product cycle concept. Sales are initially slow, progress rapidly, then become stable and eventually decline. A similar process can be observed in tourism areas. At first, places encounter restrictions related to access, facilities, information and knowledge, and visitor numbers are not significant. After this stage, facilities will become available and knowledge of the area will spread, resulting in increased visitor numbers. The popularity of the destination will develop quickly through marketing campaigns and further investments supporting the provision of hospitality services to tourists. None the less, if destinations reach levels of resource use exceeding their carrying capacity, they can enter a phase of decline and experience a decrease in visitor numbers. From less developed to developed island areas, the challenge of sustainability can be declined in diverse ways, with different categories of stakeholders involved and tourism segments attracted in that area, as it is represented in the case studies the contributors to this book on islands and tourism have analysed and deepened.

In the following pages, a critical analysis of the literature is conducted, and linkages to the main themes of the book are correspondingly discussed.

Carrying Capacity theorization follows the spread of the environmental consciousness worldwide in the second half of the 20th century. The conservation of the environment for benefiting also next generations is regarded in the aspects of economy, society, tourist satisfaction and the natural environment as well (UNWTO, 2004). Environmental concern in sustainable development characterizes the period of the late 1980s (Ahn *et al.*, 2002), since the World Commission on Economic Development (WCED) stated the most recognized definition that associates sustainability and development that 'seeks to meet the needs and aspirations of the present without compromising the ability to meet those of the future' (WCED, 1987, p. 34). Known as the Brundtland Report, it disseminated the popularity of the concept highlighting major concerns for global agendas, and still having the power to influence the debate (Telfer, 2012) on long-term viability of good development for the quality of natural and human resources. Tourism discussion adopted, then, the sustainable development paradigm. Ecotourism, responsible tourism, accessible tourism, community-based tourism and the like, were considered suitable forms of developing tourism in a sustainable way. Growth of evolution of viable strategies and managerial practices in the perspective of sustainable development gave the stage to public planning and governmental regulations, corporate social responsibility adoption, environmental certifications and codes of conduct (Telfer, 2012). A holistic vision of a tourism area includes all the components of tourism sustainability referred also to the various stakeholders, including host communities and visitors, and the management of natural and social resources.

While the diffusion of the Brundtland Report can be regarded as a milestone in the development of sustainable concerns about the environment, it also represents the affirmation of a human-centred approach to sustainable development (Butler, 1999; Fennel, 2006; Telfer, 2012). Limits to economic growth are determined by the combination of both the natural sphere and human activities. In this perspective the capacity of a tourism area to carry on a managed and monitored growth in tourism flows and services, i.e. Carrying Capacity, and the limitation accepted by primary stakeholders in the development of an area, i.e. Limits of Acceptable Change or Use, appears strictly related to the stage of development in which a geographic area can be classified, from involvement to decline, i.e. TALC. The ability of public and private managers to planning and monitoring a destination system through managerial practices such as K and LAC, and the approach of stakeholders, that are also citizens and visitors, can influence the direction of the tourism area curve of life, towards floridity or degradation. Respectively, in the first hypothesis the curve will see an increase of tourism flows during time; in the second hypothesis the curve of life of a tourism resort will see a gradual decrease of tourism flows and related tourism services over time.

Sustainable tourism management in destinations can include different tools and good practices, with the involvement of public and private interests to the environment, and international organizations involved in the preservation of the natural, cultural and historical heritage. The possible positive effects and impacts of the management of tourism in a sustainable perspective can be shown through a summarized graphic representation in Fig. 1.1. Besides the environmental advantages, also the economic and social benefit that can affect the level of employment of human resources and profitability of businesses (McCool *et al.*, 2001; Choi and Sirakaya, 2006; Zuzana and Zuzana, 2015) is highlighted; in other words the wealth of a community in a managed and monitored environment.

In Fig. 1.1, the role of the managerial tools derived by the K theory and LAC to develop strategies of a tourism area is evident. The relationship between planning through a managerial approach, the variegated measures to be adopted, and the two theories considered, are briefly discussed to provide a practical framework of reference.

The management and monitoring of tourism impacts on the social and natural environment can be considered as independent variables and have the potential to maintain the linkages between the multifaceted aspect of the core theme of sustainability and its conjugation in destinations worldwide.

The UN World Tourism Organization defines carrying capacity as 'the maximum number of people that may visit a tourist destination at the same time without causing destruction of the physical, economic or socio-cultural environment and an unacceptable decrease in the quality of the tourist satisfaction' (UNWTO, 1992, p. 23). The well-known ambits of sustainability, which are environment, economy, community and tourist satisfaction (UNWTO, 2004), are highlighted. Typically, K analysis is used for destination tourism planning as an impulse to estimate quantitatively the number of tourists an area can accommodate. This characteristic is appropriate to discuss sustainable tourism development, as well as in the context of island tourism and development perspectives considered in the island case study of the current chapter.

Farrell (1992) and Holden (2008) refer to four typologies of capacity to be managed in tourism planning: economic, psychological, environmental and social. First, the economic capacity of the tourism industry is crucial to the destination equilibrium. It is intended to provide a limitation in the development of tourism services to avoid a disequilibrium determined by a possible dependency of the destination economy upon tourism. This issue is particularly evident in destinations affected by seasonality of tourism and concentration of tourism flows in time and

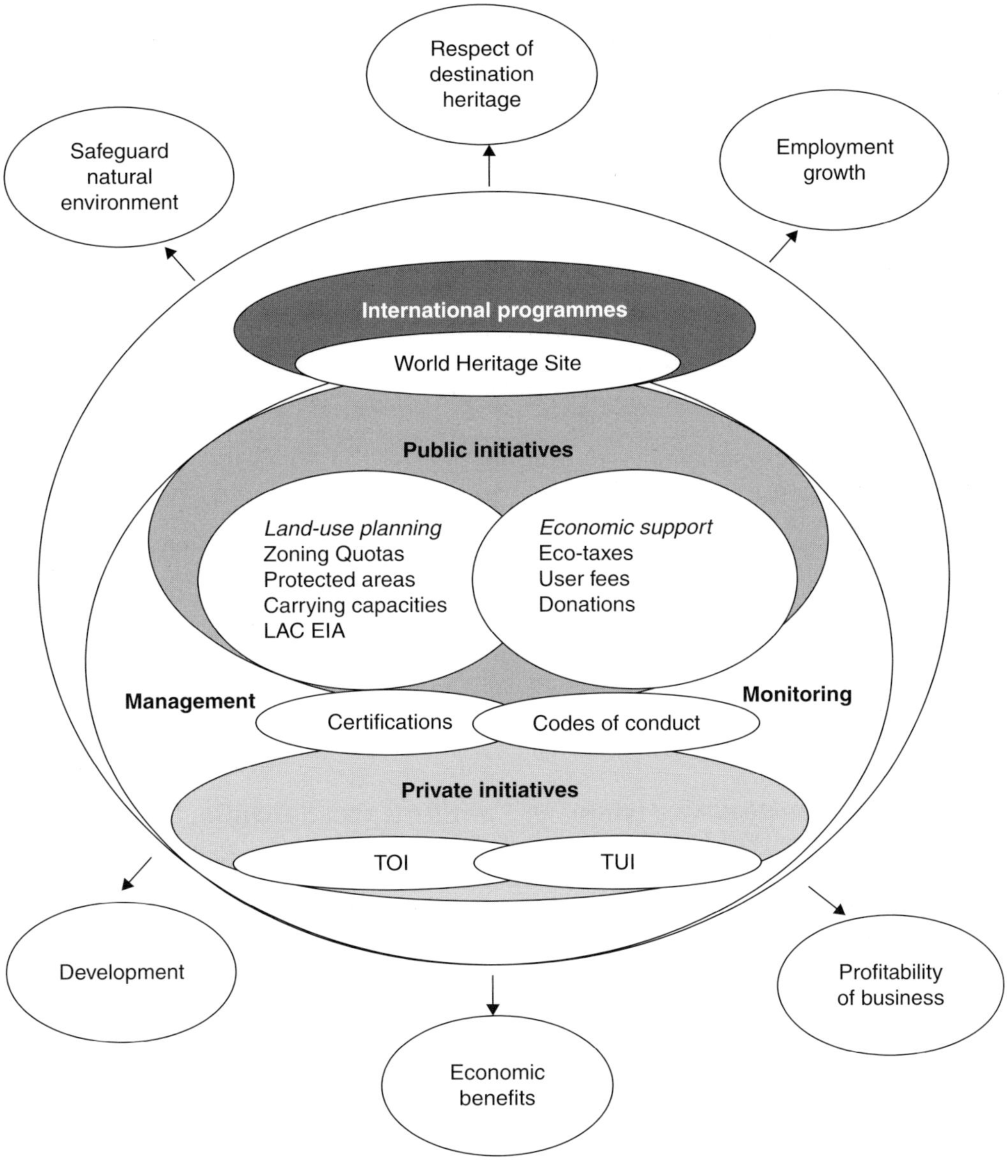

Fig. 1.1. Sustainable tourism management: frameworks and tools (Modica, 2015, p. 40).

space. Second, the psychological capacity of a destination is linked to the level of satisfaction expressed by visitors, which can diminish in case of disequilibrium in some aspects of capacity the tourism area can carry. Third, environmental capacity is easier to perceive, as damages in this aspect are immediately detectable. Therefore, also destination stakeholders will straightforwardly realize the consequences of disruptive practices. Finally, the social carrying capacity represents an essential feature, as it consists of the reaction of the local community to tourism.

A fundamental assumption is that the four different types in which K can be expressed and measured are not simultaneously essential for destination equilibrium in the short-term. An increased number of tourists in a natural area, that needs to be protected, can cause a modification in the natural environment, but the management and monitoring of the changes can direct actions to the preservation of that environment and the renewed equilibrium. In the medium- and long-term perspective, equilibriums are essentials for the conservation of a tourism area as a destination. Conversely, long-term lack of equilibriums in the social, environmental and economic aspects of sustainability can preclude the existence of a territory for tourism activities.

Tourism literature provides operationalization of destination carrying capacity with LAC, through a cooperative approach among different categories of stakeholder that can be involved in destination tourism development and planning decision-making, favouring the conditions for a dynamic equilibrium. Essentially, the LAC framework allows to assess the impacts of activities on tourism destination areas (Mbaiwa, 2015), regarded as complex systems including different dimensions managed using monitoring tools to evaluate policies' effectiveness in the social, economic and environmental aspects. As McCool highlights (2012), LAC is both a step-based framework (see Table 1.2, incorporating LAC steps adopted in the case study island) to define management planning and actions required, and a destination issue for debate, i.e. a discussion over the suitability and tolerability of varied situations. This second view has been used in the following case study destination.

1.3 A Case Study of the Island of Sant'Antioco, Sardinia, Italy

This case study refers to a typical example of a small island in the Mediterranean Sea with a potential increase in the development rate, including sustainable tourism: the island of Sant'Antioco, located off the south-western coast of the island of Sardinia, Italy. The selection of this island has been also influenced by its morphological feature: an island proximal to another island. Sant'Antioco is situated in the sea surrounding Sardinia and is linked to it, both far from the peninsular territory of Italy, i.e. an island of an island.

The case study focuses on a relevant category of stakeholders in a sustainable destination development: local entrepreneurs, a small group of which, working within the tourism and related sectors, such as hotels, B&B, restaurants, agri-food, fishery and transportation, was involved. A first approach to local entrepreneurs in this area of Sardinia (Sulcis) was facilitated, previously to this study, thanks to the participation of a university consortium research project 'Beautiful mines' submitted to the Sardinian regional government for tourism planning in this territory. Currently the research project is being subjected to evaluation. Lack of funding explains a general qualitative approach applied to the in-field activities, together with the difficulty in the retrieval of specific touristic data of the island.

The case study focuses on the territory of the Municipality of Sant'Antioco (see map, Fig. 1.2).

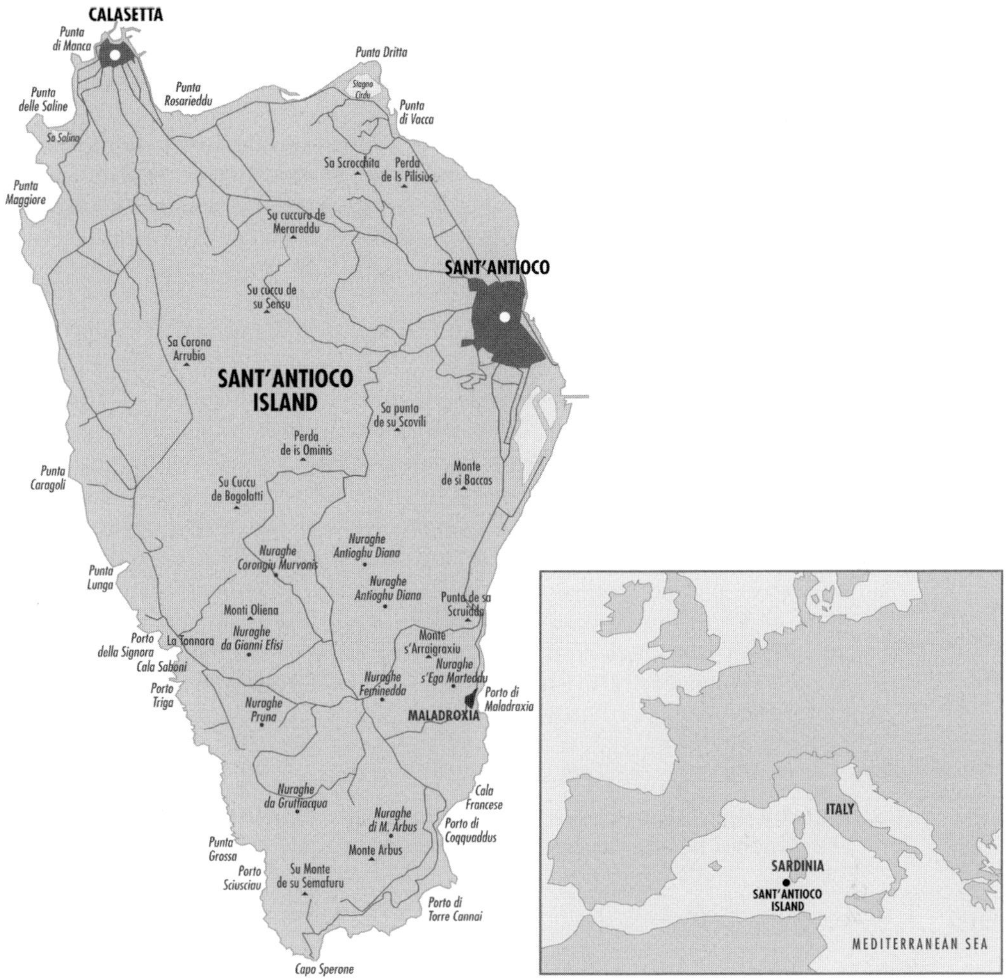

Fig. 1.2. Map of the island of Sant'Antioco, Sardinia.

1.3.1 Background

According to the interviews addressed to entrepreneurs in the preliminary test of the case study, the island of Sant'Antioco, in terms of tourism development, cannot be easily included in a specific phase of life of the Butler TALC framework. Respondents referred to both development and stagnation or consolidation and decline. This appears coherent to a socio-economic and environmental situation in the area of Sulcis in that part of Sardinia, characterized in the past by mines and industries no more in use.

The destination is linked by road through a 3 km isthmus and a bridge to the mainland island of Sardinia. It comprises two municipalities, which are administered by the regional government, and has a total area of about 109 km^2. The population is predominantly indigenous, and the population density is approximately 132 inhabitants/km^2. A rough and wild coastline, offering indented promontories on the southern and western coasts and sandy beaches on the eastern and north-western side, characterizes the island. The Mediterranean climate, with mild winters and hot and dry summers, makes the destination attractive during most part of the year, and frequent winds represent a strong point, boosting watersports, such as kite surfing and windsurfing.

With regard to the economy of the island, fishing is still an important economic activity and Carignano wine production boasts an ancient heritage, which actually represents a distinctive trait of the area. Tourism in the destination is currently becoming a key economic driver, although the industry is affected by seasonality, with a tourism season spanning approximately 5 months. This seasonality contributes to the growth of unemployment rates.

The destination is home to many tourist attractions, with nature-based attractions being significantly important. The sun and the sea are the destination's main attractions. Other natural strengths include European protected sites, and historical locations that are also significant. In addition, there are cultural events such as religious and non-religious celebrations that take place throughout the year.

The destination is home to primary, secondary and high schools, including a vocational training centre focused on tourism services. Regarding transportation to and within the destination, there are numerous provincial and national roads, railways in the nearest municipalities in the mainland island, marinas and ports, and an international airport 70 km from the island of Sant'Antioco. The destination has hotels and other services for tourists, including 14 hotels with a total number of 974 beds, while 1150 more beds are offered by other accommodation providers. Unemployment is a major social issue. From a touristic point of view the destination is not yet well known, with a potential in spreading its image, linked to southern Sardinia, both within the national and international tourism market.

The principal features are summarized below in Table 1.1.

Table 1.1. Overview of the case study island.

General features	Area	About 109 km^2
	Morphology	Rough and wild coastline, indented promontories and sandy beaches
	Population	Predominantly indigenous Population density is 132 inhabitants/km^2
	Climate	Mediterranean
	Economy	Based principally on the primary and service sectors
Attributes	Attractions	Wide variety of natural, historical, archaeological and cultural attractions. Strengths are the sea and the beaches
	Infrastructure	Developing
	Accessibility	Accessibility in a broad sense: medium-high
	Human resource	Average level of education
	Price	Low-medium
	Image	Low sub-national image

1.3.2 Case study design and methods

A qualitative methodology was used to develop the case study. Two focused visits to the island site, individual meetings with key island stakeholders, semi-structured interviews and a stakeholder focus group discussion allowed the collection of information and data. An academic member and an international expert on participant observation, taking field notes of the island destination context, conducted the visits to the island. Semi-structured interviews were, initially, addressed to a small group of representatives of the tourism entrepreneurs with the aim of acquiring general information on the tourism sector in the destination and specific information linked to the theories considered in this chapter. In addition, the entrepreneurial approach to tourism development in the island was also investigated.

As highlighted by Morgan (1996) and in line with its nature based on the group interaction, a focus group discussion was conducted, to debate with a group of 14 entrepreneurs,[1] from different backgrounds, about the sustainable tourism development of Sant'Antioco within the framework of the LAC theory. The representatives of the private sector, formed through snowball sampling (Bryman and Bell, 2015), comprised different businesses directly or indirectly related to tourism and services required by tourists visiting the island. A strong knowledge of the island, its strengths and weaknesses, and a cooperative attitude were needed to build up a positive and proactive approach during the discussion on tourism in the island, its development and the possible changes in the perspective of sustainable practices and principles.

With the declared intention to deepen the potential of LAC in stimulating users to consider their preferences, values and beliefs and identify agreement on outcomes of acceptable change, the LAC steps listed in Table 1.2 were specifically faced within the focus group. The choice for these specific LAC steps, within the general framework of LAC steps theory, derived from the availability of information retrievable in the context of the focus group discussion. Content and analysis are highlighted below.

Table 1.2. Focus group discussion in terms of sustainable tourism development (LAC steps adapted from McCool, 2012).

LAC steps selected	Purpose
Identify area concern and issues	Stakeholder meeting to identify areas of interests, to be developed or conserved
Define and describe opportunity classes	The area is divided into classes with different opportunity of development
Inventory resource and social conditions	This step aims to acquire complete knowledge of destination resources
Identify alternative opportunities class allocation	Based on steps 1 and 3, this step evaluates alternative use of opportunity classes, composing contended interests and values
Identify management actions for each alternative	Evaluation of costs, investments and benefits for each alternative
Evaluation and selection of a preferred alternative	Identification of the preferred alternative

In accordance with focus group discussion literature (Stewart and Shamdasani, 2015), the group was limited in the number of individuals, but an over recruiting of 35% was done to assure the level of participation required (Morgan, 1996; Wilkinson, 2004). The focus group met once, consequently data and theoretical saturation has not been reached; future research needs to be developed. The focus group was planned to last a maximum of 2 hours, with two leading figures from the research team, the focus coordinator and an expert assistant (Krueger and Casey, 2000). To facilitate the entrepreneur's participation and the meeting, it was decided to select the venue in a hotel in the island context.

1.3.3 Results

Interviews addressed face to face to a selected number of entrepreneurs from the island of Sant'Antioco allowed focalizing the core aspect on which the following discussion of the focus group centred. A lack of equilibrium in tourism flows had emerged from the interviews and the respondents feeling about the TALC stage of the resort area evidenced that probably the trend evolution will be towards stagnation or decline. The participants referred that the peak period in tourism flows is concentrated prevalently in summer weeks, more precisely in the month of August, while the focus group discussion also highlighted the related serious managerial concerns. This is consistent to the seasonality of tourism flows to the island of Sardinia and, in general, to the Mediterranean islands. However, in this case, with a more relevant peak deriving, among other factors, from the composition of tourist arrivals and the nights spent in the island prevalently by Italian guests (ISTAT, 2009–2014).

Based on this general assumption of lack of equilibrium in tourism flows, the LAC focus group centred on these following LAC steps, on which the entrepreneur group was invited to express individual information and opinions, and collegial thoughts:

- Priority area that requires attention related to the hospitality industry and seasonality of tourism flows: expose principal management concerns;
- Opportunity classes: areas/zones with different natural, social and managerial problems (e.g. three zones: beaches, city centre, inland). Choose one zone to go ahead with the other steps;
- Inventory resources: infrastructures, biophysical attributes and social attributes;
- Management of different alternatives;
- Allocation of costs, investments, benefits to the different development alternatives; and
- Evaluation and selection of a preferred alternative.

Initially, the official provincial data of tourism flows for the period 2009–2014 were set out to make the entrepreneur group become familiar with the priority areas emerged during the preliminary phase of interviews: the hospitality industry and seasonality of tourism flows to the island. The discussion highlighted the main management concerns. Visitors to the island are prevalently attracted by the well-known '3 S' paradigm: Sun, Sand and Sea. In addition, the Municipality of Sant'Antioco has only 4 km of sandy beaches available for tourism. Extreme seasonality and human pressure in the peak period influences the quality and quantity of tourism supply throughout the year. High fixed costs of the hotel industry do not allow the expansion of opening periods. A strong regional and sub-regional regulation in the utilization of coastal resources also characterizes the island of Sardinia, and 15% of entrepreneurs argued on the limitation for private investments. None the less, another 15% of them marked the need for requalification of the existing tourism facilities to increase tourist arrivals in off-peak periods. Moreover, the overall hospitality shows numerous aspects for improvement. And finally, the need for better coordination, planning and promotion of the destination system, also involving the public sector and other relevant stakeholders was highlighted by 38% of them.

Secondly, the following LAC step proposed to the group discussion, focused on the resort area and different classes linked to the related natural, social and managerial issues, was illustrated. Three potential zones were suggested: beaches, municipality centre and inland area. The thorough knowledge of the destination on behalf of the group of entrepreneurs allowed to specify the following opportunity classes: sea, principal inhabited area, lagoon and canals, subterranean city and catacombs, inland natural territory. The debate appeared, immediately, profitable in terms of ideas and information on the island and the richness of environments. Different opinions tended to aggregate three of these zones, specifically the multifaceted seafront comprised of beaches, coastline, the marina and industrial port, the lagoon area and the related industries, and the vast area of the city centre with its historical and archaeological interest. Upon the focus coordinator's invitation to choose one of these potential destination opportunity classes on which to continue the LAC discourse, the abundant majority of the entrepreneurs chose the articulated seafront.

Thirdly, the entrepreneurs were invited to make a general inventory of resources related to the seafront in terms of infrastructures, attractions, businesses, and biophysical and social attributes. A participated contribution of the entrepreneurs group to the third LAC step allowed detecting and classifying the island resources pertaining to the overall seafront in the Municipality of Sant'Antioco, as follows: community, businesses, infrastructures and natural resources. The social community comprises the indigenous population and other communities, all related in some ways to the sea: an important community of fishermen, the most developed in Sardinia, with also a group from Sicily; a community of military forces in the port, and the Association of Italian Sailors. The entrepreneurs were also proud that their saint, Antioco Martyr, which is also the Patron Saint of Sardinia, arrived in the past from the sea; they named him as part of the ancient community coming from the neighbouring countries. A rich variety of businesses and activities are related to this opportunity class, such as fish farms, farming and processing of fish products, fishing tourism, shipwrights, rush fishing baskets and lobster pots craftsmanship, sea silk, shipbuilding maintenance and boat storage, boat sales, vine and Carignano wine, diving, water sports, sailing and rowing schools. Infrastructures were numbered, and among them a huge marina, with touristic and industrial infrastructures, saltpans, all the tourism facilities including hotels and restaurants, an international airport near the capital of Sardinia, Cagliari, and a pilot project for a seaplane base, and the museum of the sea. With regard to natural resources, about six Sites of Community Interest (SIC) regulated by the European Union and about three Zones for Special Protection (ZPS) were named, and the preservation of the dune system is of particular concern. None the less, the entrepreneurs highlighted the lack of a marine protected area for the preservation of this unique habitat, for example, the presence of the 'pinna nobilis' from which sea silk is made, an autochthon orchid and other species are equally of great interest.

Steps four and five of the Sant'Antioco island LAC exercise were then taken into consideration. Different management alternatives of the chosen sea area were proposed and discussed by the group of entrepreneurs, together with an estimate of the costs and benefits for the different alternatives: increase of moorings; protected coastal area; fish farms, lagoon and pond; water sports; saltpans. The increase in moorings, as a development area deserving attention, was stressed by one of the participants with a long experience on port management. He evidenced the sustainability of floating docks with low environmental impact and an increase in employment, limited investments and the desirable inclusion in the Sulcis Regional Master Plan. The increase of moorings seemed to be of interest for the stakeholder working group within the LAC exercise. The preservation of the seafront through the creation of a protected area was then considered. Business activities, including fishing, are strictly related to this part of the municipality, hence the reaction of the fishermen community could be of concern. The economic tourism development of the marina front was considered in terms of benefits that would derive from this type of intervention, as the focus-group participants estimated that the costs of the endeavour, with reference to the local context, would however be contained. The debate then focused on the relation existing between the fish farms and the lagoon and pond facing the town. Farming and processing of fish products are important industries for the community and conflicting interests could emerge from a new and diverse utilization of them; investments on education of the human resources involved in this would change the approach of the fishermen's community, while the high environmental value of the lagoon and the pond were also pointed out. Water sports were then discussed, as important international events are linked to this, thanks also to the strong winds that characterize the island. This appeared to be considered another asset to be strengthened as the infrastructures and investments needed are very limited and have a considerably low impact on the natural environment. However, also on this front from the discussion the risk for a possible social conflict emerged between the important community of fishermen and the sports people involved. Nevertheless, the potential of water sports in reducing the seasonality of the tourism flows, favouring the hospitality industry occupancy rates, was stressed. Finally, saltpans were considered thoroughly. This topic in reality absorbed the majority of the time dedicated to these two LAC steps. The entrepreneurs emphasized the potential of this industry, in fact, their salt in terms of wellness and medical use is of great value. Currently, saltpans are owned and managed for the industrial commercialization of the extracted salt by a southern Italian company. The same company develops all the other uses of this salt in southern Italy. The entrepreneurs are not informed about the relationship existing between their community and the saltpan company. However, due to previous industrial activities, such as a disused power station, the saltpan area is also currently an environmental concern. Numerous hypotheses for potential alternative uses for the premises of dismantled industries or establishments were taken into consideration ranging from bird watching to spas or wellness centres to be established, production of cosmetics, also in cooperation with the regional medical governmental structures, etc.

Finally, LAC step number six was proposed concerning the evaluation and selection of a preferred management alternative within the ones listed, in the perspective of a sustainable development of the island. At the end of the group discussion, the saltpans' development in a tourism perspective and the institution of a protected area remained the most appreciated management alternatives. However, the saltpan option was evaluated as futuristic; varied circumstances prove that they are not yet prepared for the economic development of the saltpan. At last, hence, the protected coastal area was the final winner, with an 85% of entrepreneurial consensus.

1.3.4 Lessons learned

The utilization, with the cooperation of a primary category of destination stakeholders (Aas *et al.*, 2005; Getz and Timur, 2005), of the LAC framework in a realm such as the Mediterranean island context has several points of interests for sustainable island destination planning, management and monitoring, which will be listed, commented and recalled further.

Stakeholder meetings, as part of the participatory approach in a sustainable development, show their effectiveness for a democratic composition of diverse individual interests. Nevertheless, in the LAC discourse with the entrepreneurs in the island of Sant'Antioco, the possibility of converging decision-making to sustainable development has been favoured by the necessity to identify changes accepted by all.

Moreover, the necessary coordinated approach for defining specific objectives in the island development has the potential to favour stakeholders' awareness over their peculiar resources.

In addition, another lesson learned is related to the practical organization of stakeholder meetings in the effort to obtain important feedback for decision-making in sustainable island tourism. Instead of organizing a general assembly with all stakeholder categories, it appeared more effective, in terms of LAC outputs, to hold separate sessions with relevant stakeholders more than a unitary decision-making process. This way it was possible to make stakeholders aware and become active in the planning process, separately through a peer confrontation procedure, which, can then contribute in a concrete and non-disruptive way to a plenary session on an overall LAC implementation.

1.4 Conclusions

This chapter focused on sustainability and island tourism development, including a case study of a Sardinian island.

The chapter started from the consideration of tourism as a possible vehicle of stable modification in the natural, social and economic environment, as the WBGU (1996) stated for mass tourism and consequences for the planet.

The unique and fragile *equilibrium* that characterizes island destinations, devoted to or developing tourism activities, gave the reason to debate on the binomial island and tourism in the perspective of sustainable development. The three pillars in which sustainable tourism is usually regarded (UNWTO, 2004; UNEP and UNWTO, 2005; ILO, 2010) are all essential to discuss equilibriums in island destinations. The managerial aspect in sustainable tourism (Miller and Twining-Ward, 2005;

Modica, 2015) contributes to achieve both heritage and landscape conservation and economic development. Carrying Capacity and Limits of Acceptable Change theoretical frameworks seemed to incorporate managerial tools suitable to investigate the application of the theme equilibrium in islands as tourism destinations. Tourism literature provides operationalization of destination carrying capacity with LAC, through a cooperative approach among different categories of stakeholder that can be involved in destination tourism development and planning decision-making, favouring the conditions for a dynamic equilibrium. As McCool (2012) highlights, LAC is both a step-based framework to define management planning and actions required, and a destination issue for debate, i.e. a discussion over the suitability and tolerability of varied situations. This second view has been used in the Sardinian case study destination.

The case study referred to the island of Sant'Antioco, located off the south-western coast of the island of Sardinia, Italy. A small group of entrepreneurs working within tourism and related sectors was involved to test the feasibility of LAC to positively debate sustainable island tourism. A general qualitative methodology was used: semi-structured interviews and a focus group discussion.

The utilization, with the cooperation of primary categories of destination stakeholders, of the LAC framework in a realm such as the Mediterranean island context revealed interest for sustainable island destination planning, management and monitoring. Based on the assessment, through interviews, of lack of equilibrium in tourism flows, the focus group went through a discussion on sustainable 'management by objectives' that characterizes the LAC process. The output of the entrepreneurial debate was the awareness raising of the stakeholders of their peculiar resources and the identification of changes acceptable by all. In the perspective of island sustainable development, within a LAC framework, the case study focus group oriented their preferred management alternative to the settlement of a marine protected area, as a value, also to spend in the tourism market.

Note

1 The authors would like to sincerely thank the entrepreneurs for their important contribution to the focus group discussion.

References

Aas, C., Ladkin, A. and Fletcher, J. (2005) Stakeholder collaboration and heritage management. *Annals of Tourism Research* 32(1), 28–48.

Ahn, B., Lee, B. and Shafer, C.S. (2002) Operationalizing sustainability in regional tourism planning: an application of the limits of acceptable change framework. *Tourism Management* 23(1), 1–15.

Baggio, R. (2008) Symptoms of complexities in a tourism system. *Tourism Analysis* 13(1), 1–20.

Beaumont, N. and Dredge, D. (2010) Local tourism governance: a comparison of three network approaches. *Journal of Sustainable Tourism* 18(1), 1–22.

Bryman, A. and Bell, E. (2015) *Business Research Methods*. Oxford University Press, Oxford, UK.

Butler, R.W. (1980) The concept of the tourist area life-cycle of evolution: implications for management of resources. *Canadian Geographer* 24(1), 5–12.

Butler, R.W. (1999) Sustainable tourism: a state-of-art-review. *Tourism Geographies* 1, 7–25.

Choi, H.C. and Sirakaya, E. (2006) Sustainability indicators for managing community tourism. *Tourism Management* 27(6), 1274–1289.

Farrell, B. (1992) Tourism as an element in sustainable development: Hana Maui. In: Smith, V.L. and Eadington, W.R. (eds) *Tourism Alternatives: Potentials and Problems in the Development of Tourism.* University of Pennsylvania Press, Philadelphia, pp. 115–132.

Faulkner, B. and Russell, R. (2001) Turbulence, chaos and complexity in tourism systems: a research direction for the new millennium. In: Faulkner, B., Moscardo, G. and Laws, E. (eds) *Tourism in the 21st Century: Lessons from Experience.* Continuum, London, pp. 328–349.

Fennell, D. (2006) *Tourism Ethics.* Channel View Publications, Clevedon, UK.

Frauman E. and Banks, S. (2011) Gateway community resident perceptions of tourism development: incorporating importance-performance analysis into a limits of acceptable change framework. *Tourism Management* 32, 128–140.

Getz, D. and Timur, S. (2005) *Stakeholder Involvement in Sustainable Tourism: balancing the voices*, 3rd edn. Elsevier, London.

Hall, M. (2006) Introduction. In: Butler, R.W. (ed.) *The Tourism Area Life Cycle. Applications and Modifications,* Vol. 1. Channel View Publications, Clevedon, UK, p. xv.

Hall, M., Gössling, S. and Scott, D. (2015) *The Routledge Handbook of Tourism and Sustainability.* Routledge, London.

Holden, A. (2008) *Environment and Tourism*, 2nd edn. Routledge, London.

ILO (International Labor Organization) (2010) Developments and challenges in the hospitality and tourism sector. Sectoral Activities Programme. Issues paper for discussion at the Global Dialogue Forum for the Hotels, Catering, Tourism Sector (23–24 November 2010).

ISTAT (Italian Institute of Statistics) (2009–2014) Capacity and occupancy of tourist accomodation establishments, 2009–2014. Available at: http://dati.istat.it (accessed 10 August 2016).

Karatzoglou, B. and Spilanis, I. (2010) Sustainable tourism in Greek islands: the integration of activity-based environmental management with a destination environmental scorecard based on the adaptive resource management paradigm. *Business Strategy and the Environment* 19(1), 26–38.

Kerstetter, D. and Bricker, K. (2012) Relationship between carrying capacity of small island tourism destinations and quality-of-life. In: Uysal, M., Perdue, R. and Sirgy, J. (eds) *Handbook of Tourism and Quality-of-Life: The Missing Links.* Springer, the Netherlands, pp. 445–462.

Krueger, R.A. and Casey, M.A. (2000) *Focus Groups: A practical guide for applied researchers*, 3rd edn. Sage, Thousand Oaks, California.

Lagiewski, M.R. (2006) The application of the TALC model: a literature survey. In: Butler, R.W. (ed.) *The Tourism Area Life Cycle,* Vol. 2. Channel View Publications, Clevedon, UK, pp. 27–50.

Larson, L.R. and Poudyal, N.C. (2012) Developing sustainable tourism through adaptive resource management: a case study of Machu Picchu, Peru. *Journal of Sustainable Tourism* 20(7), 1–22.

Lusseau, D. (2008) Carrying capacity. In: Lück, M. (ed.) *The Encyclopedia of Tourism and Recreation in Marine Environments.* CABI Publishing, Wallingford, UK, pp. 84–86.

Mbaiwa, J. (2015) Limits of acceptable change (LAC). In: Cater, C., Garrod, B. and Low, T. (eds) *The Encyclopedia of Sustainable Tourism.* CABI Publishing, Wallingford, UK, p. 315.

McCook, L.J., Ayling, T., Cappo, M., Choat, J.H., Evans, R.D., De Freitas, D.M. and Williamson, D.H. (2010) Adaptive management of the Great Barrier Reef: a globally significant demonstration of the benefits of networks of marine reserves. *Proceedings of the National Academy of Sciences of the United States of America* 107(43), 18278–18285.

McCool, S.F. (2012) Limits of acceptable change and tourism. In: Fennel, D. and Holden, A. (eds) *The Routledge Handbook of Tourism and the Environment.* Routledge, London, pp. 285–298.

McCool, S.F., Moisey, R.N. and Nickerson, N.P. (2001) What should tourism sustain? The disconnect with industry perceptions of useful indicators. *Journal of Travel Research* 40(2), 124–131.

McMahon, E.T. (1999) Gateway communities. *Planning Commisioners Journal* 34, 6–7.

Miller, G. and Twining-Ward, L. (2005) *Monitoring for a Sustainable Tourism Transition: the challenge of developing and using indicators.* CABI Publishing, Wallingford, UK.

Modica, P. (2015) *Sustainable Tourism Mangement and Monitoring. Destination, Business and Stakeholder Perspectives*. Franco Angeli, Milan.

Morgan, D.L. (1996) Focus groups. *Annual Review of Sociology* 22, 129–152.

Pechlaner, H., Tallinucci, V., Abfalter, D. and Rienzner, H. (2003) Networking for small island destinations – the case of Elba. In: Frew, A.J., Hitz, M. and O'Connor, P. (eds) *Information and Communication Technologies in Tourism* Springer, Vienna, pp. 105–114.

Plummer, R. and Fennell, D. (2009) Managing protected areas for sustainable tourism: prospects for adaptive co-management. *Journal of Sustainable Tourism* 17(2), 149–168.

Schianetz, K., Kavanagh, L. and Lockington, D. (2007) The learning tourism destination: The potential of a learning organisation approach for improving the sustainability of tourism destinations. *Tourism Management* 28(6), 1485–1496.

Scott, N., Baggio, R. and Cooper, C. (2008) *Network Analysis and Tourism.* Channel View Publications, Clevedon, UK.

Stewart, D.W. and Shamdasani, P.N. (2015) *Focus Groups: Theory and practice.* Sage, Thousand Oaks, California.

Telfer, D. (2012) The Brundtland Report (Our Common Future) and tourism. In: Holden, A. and Fennell, D. (eds) *The Routledge Handbook of Tourism and the Environment.* Routledge, London.

UNEP and UNWTO (United Nations Environment Programme, and World Tourism Organization) (2005) *Making Tourism More Sustainable. A Guide for Policy Makers.* United Nations Environment Program (UNEP) and World Tourism Organization (WTO), Paris and Madrid, pp. 11–12.

UNWTO (United Nations World Tourism Organization) (1992) *Tourism Carrying Capacity.* Report on the senior-level expert group meeting held in Paris, June 1990. UNWTO, Madrid.

UNWTO (United Nations World Tourism Organization) (2004) *Indicators of Sustainable Development for Tourism Destinations.* UNWTO, Madrid.

UNWTO (United Nations World Tourism Organization) (2007) *A Practical Guide to Tourism Destination Management.* UNWTO, Madrid.

WBGU (1996) *World in Transition: the research challenge.* Annual Report 1996, WBGU – German Advisory Council on Global Change. Springer-Verlag, Berlin.

WCED (World Commission on Environment and Development) (1987) *Our Common Future.* Oxford University Press, Oxford, UK.

Wilkinson, S. (2004) Focus group research. In: Silverman, D. (ed.) *Qualitative Research: Theory, method, and practice.* Sage, Thousand Oaks, California, pp. 177–199.

Yang, C. and Sun, X. (2013) Evolution of a tourism destination complex system. In: Ryan, C. and Huan, S. (eds) *Tourism in China: Destinations, Planning and Experiences.* Channel View Publications, Clevedon, UK, pp. 146–168.

Zuzana, J. and Zuzana, L. (2015) Monitoring system of sustainable development in cultural and mountain tourism destinations. *Journal of Competitiveness* 7(1), 35–52.

2 Ecotourism and Environmental Management: A Case Study of a Partnership for Conservation in Fiji

KELLY S. BRICKER[1]* AND DEBORAH L. KERSTETTER[2]

[1]University of Utah, USA; [2]The Pennsylvania State University, USA

2.1 Introduction

The purpose of this chapter is to provide an overview of environmental management in Fiji. This is accomplished through a review of the current situation in Fiji as well as the presentation of a case study of Rivers Fiji, an ecotourism company that partnered with native landowners to facilitate a lease for conservation of the Upper Navua River.

2.2 Fiji: A South Pacific Island Country

Fiji lies nearly 1900 miles north of Sydney, Australia and just over 3000 miles southwest of Hawaii. It is comprised of over 330 islands, the largest of which is Viti Levu, home to the capital city, Suva, and approximately 70% of the nation's 887,000 residents (World Bank, 2015). Approximately 57% of Fiji's residents are indigenous Fijians and 38% are Indian. Of indigenous Fijians, 55% live in rural areas (FIBS: Fiji Islands Bureau of Statistics, 2015) and nearly all are heavily dependent on farming for their subsistence and livelihood (WCS: Wildlife Conservation Society, 2009).

Fiji is one of the most developed countries in the South Pacific, with forest, mineral and fish resources. Its major sources of foreign exchange are sugar exports, remittances from those working abroad and tourism, which has experienced significant growth (692,630 visitors in 2014) (FIBS, 2015). Fiji is also an unstable country. Since 1970 there have been four political coups d'état and two abrogated constitutions. This instability has placed increased pressure on natural resources, due to the 'increased threat of extractive industries, erosion of traditional governance, and legal instability along ethnic lines' (Govan *et al.*, 2012, p. 3).

*E-mail: kelly.bricker@health.utah.edu

2.2.1 Environmental management in Fiji

During 2014 an interim government held the first credible election since the coup d'état of 2006. The election was considered a turning point for the country, providing stability and new direction. During the same year the interim government endorsed a green growth framework (GGF) for the country – 'Restoring the Balance in Development that is Sustainable for Our Future' (Republic of Fiji, 2014). The GGF recognized that key stakeholders should engage in a participatory process that 'fosters harmony between the three pillars' of sustainability (i.e. environment, social, economic) and contributes to 'Fiji's future economic development' (Republic of Fiji, 2014, p. 16). To ensure that the many types of environmental, social and economic factors be considered in a participatory process, the GGF included a list of the three pillars of sustainability as well as focal areas within them: Environment (i.e. building resilience to climate change and disasters, waste management, sustainable island and ocean resources); Social (i.e. inclusive social development, food security, freshwater resources and sanitation management); and Economic (i.e. energy security, sustainable transportation, technology innovation, greening tourism and manufacturing industries) (Republic of Fiji, 2014, pp. 16–17).

Presently, a multi-stakeholder panel comprised of representatives from government, private sector, non-governmental organizations (NGOs), academic and other key interest groups is leading the implementation of the GGF. The panel, chaired by Fiji's prime minister, is expected to provide a platform to strengthen partnerships between the various stakeholders pursuing sustainability initiatives (Republic of Fiji, 2014).

In addition to endorsing the GGF, the Fijian government has also signed and ratified four multi-lateral environmental agreements that established and/or supported management of protected areas in Fiji (Clarke and Gillespie, 2008, p. 1): 1971 Convention on Wetlands of International Importance, 1972 Convention Concerning the Protection of the World Cultural and National Heritage, 1986 Convention on the Protection of Natural Resources and the Environment of the South Pacific, and 1992 Convention on Biological Diversity. These treaties have committed Fiji to 'protect[ing] its natural and cultural heritage and to conserve[ing] and sustainably manage[ing] its biological diversity using appropriate means, including the establishment and management of protected areas' (Clarke and Gillespie, 2008, p. 1). Examples of how Fiji has managed its protected areas follow.

2.2.2 Protected areas in Fiji

Fiji's environmental protection policies and laws vary immensely, are managed by several ministries and departments and, because they have different legal status, have different levels of protection under law. While Fiji does not have any dedicated protected area legislation (Clarke and Gillespie, 2008), the government has introduced statutory mechanisms whose key features focus on land acquisition and conservation and empowering governmental units to conserve and protect endangered species (Table 2.1).

Fiji has also adopted the National Biodiversity Strategy Action Plan (NBSAP), which emphasizes the importance of terrestrial biodiversity conservation: 'the conservation and sustainable management of Fiji's natural forests is the single most important means of conserving the vast majority of Fiji's endemic fauna and flora'

Table 2.1. Legal options for the establishment of protected areas in Fiji (Clarke and Gillespie, 2008, p. 2).

Legislation	Key features	Sections
Native Lands Act	Recognizes and maintains communal ownership of native lands	3
Native (i Taukei) Land Trust Act	Establishes the Native (i Taukei) Land and Trust Board (TLTB), and empowers the Board to: • Enter into leases and licences on behalf of native landowners; • Place conditions on the use of leased land; and • Declare certain areas to be native reserves.	7–9 10
National Trust of Fiji Act	Establishes the National Trust of Fiji, and empowers the trust to: • Acquire and manage property of natural or cultural heritage significance; • Create by-laws for the regulation and protection of Trust properties; • Enter into heritage conservation covenants with landowners; and • Declare national heritage areas.	4 18 10 20
Forestry Decree	Empowers the Ministry of Environment for Forests to: • Declare multiple use forest reserves; and • Declare strict nature reserves.	7–8 7–8
Land Conservation and Improvement Act	Establishes the Land Conservation Board and empowers the board to make conservation, closing and work orders.	3 7–9
Water Supply Act	Empowers the Minister to declare protected water catchment areas.	4
Bird and Game Protection Act	Prohibits the killing, wounding, or taking of protected bird species and regulates hunting of game species.	3–6
Endangered and Protected Species Act	Prohibits the possession, sale, display or trade of listed endangered and protected species.	23
Environmental Management Act	Requires environmental impact assessment of development activities that are likely to have a significant impact on the environment, including existing or proposed protected areas	2, 27

(Ministry of Fisheries and Forestry, 2007, p. 4). Conservation focal areas include: community support – awareness, involvement and ownership, improving our knowledge, developing protected areas, species conservation, management of invasive species, and capacity building and strengthening (for specific information about these focal areas access: http://theredddesk.org/sites/default/files/national_biodiversity_strategy_and_action_plan_2007_0.pdf).

The NBSAP also includes the Convention on Biodiversity's (CBD) definition of a protected area: 'a geographic area, which is designated or regulated and managed to achieve long-term conservation of nature with associated ecosystem services and cultural values' (IUCN: International Union for the Conservation of Nature, 2008, p. 1). In Fiji, the designation status and the regulatory mechanisms associated with protected areas vary (see Table 2.1).

Despite the fact that Fiji has no dedicated legislation devoted to protecting environmental areas, its adoption of the NBSAP and acceptance of assistance from international and local NGOs has moved some conservation efforts forward (Vukikomoala *et al.*, 2012). Following are examples of these efforts (modified from Vukikomoala *et al.*, 2012, pp. 25–26).

1. A collaboration between various governmental units to assess Sovi Basin management plans has led to legislation, resource inventories and production of awareness materials.
2. Assessments by The National Protected Area Committee of the effectiveness of community-based marine management strategies that account for the amount of marine habitat 'effectively' protected in Fiji (Mills *et al.*, 2011).
3. The Wildlife Conservation Society (WCS), an NGO based in Fiji, has actively supported Fiji's efforts to implement the NBSAP.
4. Academic institutions, NGOs and resource owners supporting Fiji's 2005 commitment to set up 30% of its marine areas as a network of ecologically representative and effectively managed marine areas within the i qoli qolis (i.e. landowning groups with water access).

2.2.3 Customary Fijian management practices

According to Govan *et al.* (2012, p. 3):

> The increasing pressure on natural resources exerted by increased monetization of the economy, moves to promote mining and other extractive industries, [erode] traditional governance and political and legal instability along ethnic lines (including 4 coups d'état and 2 abrogated constitutions since independence), [and] represent threats demanding an increased attention to legal and institutional support of ICCAs. Terrestrial conservation is particularly neglected and under threat from potential land reforms and development.

Like other independent Pacific island countries, Fiji is unique with respect to land ownership. Most of it is under i Taukei (i.e. native land) or customary ownership (Govan *et al.*, 2012). Customary ownership of land and natural resources is acknowledged through various laws. The customary land tenure system was established in two pieces of legislation, the i Taukei Lands Act (TLA) and the i Taukei Lands Trust Act (TLTA). Section 3 of the TLA states that all i Taukei lands are to be held by indigenous people according to native customs and traditions (Vukikomoala *et al.*, 2012). Section 4 of the same act identifies the i Taukei Land Commission (TLC), formed in 1880, as the entity that registers i Taukei lands, identifies ownership of such lands, and classifies customary roles and migration records of communal units of land. The TLC can also set aside sites as native reserves (native lands that will not be subjected to a lease) (Vukikomoala *et al.*, 2012). A related entity, the i Taukei Fisheries Commission (TFC), is responsible for maintaining a record of the mapped and delineated boundary lines of 385 marine and 25 freshwater i qoli qoli areas. Approximately 300,000 i Taukei villagers are dependent on these areas for their livelihood (Aalbersberg *et al.*, 2005; Clarke and Jupiter, 2010).

The Environment Management Act (EMA) (Section 2) establishes the relationship between indigenous Fijians and their ancestral lands, waters, sites, sacred areas and other areas of national importance (Vukikomoala *et al.*, 2012). For hundreds of years, Fijian society has regulated resource use of marine and terrestrial environments through customary management practices that have typically included temporary harvesting closures, access restrictions, seasonal bans and catch limits (Johannes, 1978; Veitayaki, 1997). In addition, 'despite the introduction of open access provisions in Fiji law, customary governance systems remain the strongest mechanism for regulating natural resource use in rural Fiji' (Clarke and Jupiter, 2010). Customary governance systems have been utilized to manage social relationships such as taboos relating to short-term events such as funerals. More recently, these customary governance practices have been adapted to achieve long-term objectives such as food security, with several hundred coastal communities establishing locally managed marine areas (LMMAs) (Govan *et al.*, 2009; Mills *et al.*, 2011). Best practices in establishing, managing and monitoring these areas are shared through the Fiji LMMA (FLMMA) network, and throughout other Pacific island nations (LMMA, 2008).

2.2.4 Indigenous Fijians' and local communities' conserved territories and areas

Traditional communities are expanding their marine conservation efforts to include freshwater and terrestrial areas within mataqali[1] koro (village) and district[2] boundaries (Walsh, 2006; Clarke and Jupiter, 2010). This makes sense, as what happens on small islands moves to the sea and impacts the health of marine resources rapidly. Yet, out of the 163 Indigenous Fijians' and local communities' conserved territories and areas (ICCAs) established in Fiji, only 14 are terrestrial, and they only encompass 380 km^2 (Vukikomoala *et al.*, 2012). Compared to the 149 Marine Protected Areas (MPAs), which cover 17,726 km^2, a very small proportion of the terrestrial areas are protected. Local communities manage most of the sites whose legal status varies. Some are nature reserves established through National Forestry legislation. Others include heritage sites owned or leased by the National Trust of Fiji; conservation leases held by individuals and businesses; and marine and coastal community declared conservation areas (Clarke and Gillespie, 2008; Govan *et al.*, 2009). The foci of their rules and policies also differ. In some cases the management plans or specific lease conditions of these areas regulates access, or dictate what gear or equipment can be utilized, or sets limits on the number and types of species individuals can catch (LMMA, 2008; Vukikomoala *et al.*, 2012). Currently, no national policy or legal framework for protected areas exists. However, in August of 2011, the government approved moving forward with the development of a policy specific to establishing a national policy.

While not a part of the FLMMA network, freshwater ICCAs are typically 'managed under the same context of customary governance' (Govan *et al.*, 2012, p. 11). This includes strategies such as seasonal harvest bans and temporary (multi-year) bans on harvesting and cutting riparian vegetation. However, when bans are lifted conservation or protection benefits degrade rapidly and are slower to recover than MPAs (Jenkins and Jupiter, 2011).

Challenges and threats to terrestrial ICCAs and other protected areas

The main threats to ICCAs are classified as political/legal, institutional, and internal and external environmental drivers (Govan *et al.*, 2012; Vukikomoala *et al.*, 2012). While ownership rights for lands are clear, the opposite situation is true with freshwater, particularly rivers. In addition, 'a considerable amount of Fiji legislation and policy does not give legal authority to communities to fully restrict access to their ICCAs as they see fit' (Vukikomoala *et al.*, 2012, p. 12). Environmental management analysts in Fiji feel there is an urgent need for new protected area legislation to be reconciled with current provisions that may overturn protected area status in the interest of national government (Clarke and Gillespie, 2008; Govan *et al.*, 2012). One example within terrestrial environments is the Mining Act, which gives the Director of Mineral Resources the power to issue prospecting licences without owner consent and to declare a site of less than 250 ha for mining if deemed important to the country (Vukikomoala *et al.*, 2012).

With respect to institutional threats to ICCAs, the primary issue is lack of resources (human, financial, technical) to adequately support community-based management. The FLMMA network has attempted to improve coordination and communication with LMMA sites so as to document which ones are self-sufficient and which ones need additional resources. Without support for implementation of these sites, there is a tendency for people to lose interest, leading to poor compliance with ICCA rules (Govan *et al.*, 2009).

Internal and external environmental drivers are also threats to ICCAs. For example, when protected area status is seasonally or temporarily lifted, extreme cases of overharvesting by external parties occur (Vukikomoala *et al.*, 2012). A second threat is natural disasters that produce extreme weather events. They have led to an increased emphasis on building resilience in the face of unpredictable climate disturbances (Jupiter *et al.*, 2012).

Threats to natural resources are generally set amongst a host of internal political issues. In Fiji these issues include: (i) an underlying fear that a reformist government could cause native resource owners to lose their rights; (ii) overlapping functions and increased competing interests of state-funded agencies responsible for the administration of native resources; (iii) the promulgation of the Land Use Decree 2010;[3] (iv) lack of political will for reform; and (v) absence of clear and concise medium- to long-term government policies on the utilization of natural resources (see Vukikomoala *et al.*, 2012, pp. 26–28, for a more complete discussion of these issues). Vukikomoala *et al.* (2012) also indicated that one of the greatest challenges for ICCAs is being able to strike a balance between development and conservation.

2.3 Ecotourism and the Upper Navua Conservation Area

For nearly 20 years, the Fijian government has advocated for and adopted an ecotourism strategy that will spread the economic benefits typically found along coastal areas to rural communities such as the highlands of Fiji (Harrison, 1999). More recently, the government has focused on ecotourism and its ability to attract tourists (Bricker, 2001). This strategy has been fairly successful (Bricker, 2001; Kerstetter and Bricker, 2009; Kerstetter *et al.*, 2010; Bricker *et al.*, 2012) with tourists and with Fijians, most of whom generally respond positively to visitors.

Following is a case study of Rivers Fiji, Ltd (RF). The case study is a prime example of how an ecotourism company can generate partnerships in natural resource conservation using a unique strategy – the development of a lease for conservation.

Fig. 2.1. Rafting on the Upper Navua River. (Photo courtesy of N.K. Bricker.)

2.3.1 The case of Rivers Fiji, Ltd

RF was established as a commercial ecotourism business offering whitewater rafting/kayaking experiences during the latter part of 1997. Its mission (RF, 1997, p. 1) then and now is:

> To enhance guests' and indigenous peoples' awareness of and appreciation for the culture and environment. Rivers Fiji provides activities that promote conservation and preservation through environmentally sensitive and socially responsible interaction with people, landscapes, and ecosystems which make the Fijian Highlands distinctive and unique.

In setting up RF, the company had three primary objectives: (i) establish an ecotourism operation that directly and comprehensively contributes to the conservation of Fiji's natural resources; (ii) bring socio-economic benefits directly to the people through employment and supporting existing locally owned and operated businesses in areas where the business operates (i.e. the Navua River); and (iii) maintain the highest of standards of operation through internationally recognized training and state-of-the art equipment (RF, 1997, pp. 2–3). How RF addressed its conservation and socio-economic objectives is described, as are the unique aspects of the partnership that was created as a mechanism for natural resource protection and community development.

A lease for conservation

In 1997, after visiting the upper segment of the Navua River (i.e. Upper Navua) located in the highlands of Viti Levu, the founders of RF asked, 'What if tourism dollars could directly contribute to the conservation of this significantly unique river corridor?' They found out they could and established the Upper Navua Conservation Area (UNCA).

The UNCA was established in 2000 after RF partnered with nine mataqali and two villages to convince commercial loggers and gravel extracting entities to release 200 m of their leases from each side of the Upper Navua river corridor. At that time, commercial loggers could legally log within 30 m of the edge of the river. This not only created a potential hazard for tourists, but also caused increased sedimentation downstream into Beqa Lagoon, a biodiversity rich area of the coastal Pacific ocean of Viti Levu.

The partnership resulted in a lease for conservation of the Upper Navua river corridor (i.e. UNCA), which represents 615 ha of land. Located in the Province of Serua on the south-central side of Viti Levu, the UNCA lies between 18°7.65′S, 177°52.00′E and between 18°7.340′S and longitude 177°58.75′E. The nearest towns are Pacific Harbour, located about 40 km away, and Navua Town, located along the lower reaches of the Navua River. All the land within the UNCA is under customary land ownership (mataqali).

In sum, the lease consists of payments to the landowners for a conservation easement within the river corridor. It is the first lease for conservation in Fiji, and possibly all of the South Pacific (Bricker, 2001). The lease also exemplifies how RF met its first objective – to contribute to the conservation of Fiji's natural resources by introducing an ecotourism operation, i.e. renewable, low impact whitewater tourism.

Unique lease structure

The land and waters within the UNCA are under the direct management of RF as per the terms and conditions of the lease with the Native (i Taukei) Land and Trust Board (TLTB). Government entities involved with the potential long-term conservation issues in the UNCA include the Department of Forestry/Commissioner of Forests, Department for the Environment and Department of Lands and Mineral Resources.

To ensure that the conservation and operation objectives of the lease were met, RF developed the UNCA Management Plan that guides conservation efforts and day-to-day river programmes. The plan addresses: training for guides (e.g. technical skills, 'leave no trace', interpretation); the number of passengers that can sustainably be managed by RF; river management meetings with mataqalis, allowable uses and permitting; and how employment structures and opportunities are facilitated with members of mataqali along the river corridor and local communities in the region.

To provide economic benefits directly to communities, RF set up a unique payment system. First, the standard conditions of the lease payments included a one-time lease payment by RF to the TLTB and to each mataqali. In addition, RF implemented a payment structure where a portion of each passenger's fee would go directly to each of the nine landowning groups and two villages supporting the conservation area. This strategy created a win/win for communities and RF. As the business grows, everyone involved in the UNCA benefits from every river trip and, hence, theoretically maintains continuous interest in and support for the UNCA.

In addition, RF supported projects (e.g. sale of rugby jerseys and kava to tourists) that generated funds for community organizations. Indirect economic benefits also extended into the surrounding communities through RFs purchase of supplies, food, punt and road drivers, road crews, contractors, mechanics and equipment.

Wetland of international importance (Ramsar site)

To bolster support for the conservation efforts within the UNCA, RF worked with several stakeholders to identify the UNCA as Fiji's first Ramsar site. The UNCA created a unique business-for-conservation paradigm and laid the groundwork necessary to help Fiji establish its first Wetland of International Importance (i.e. Ramsar site). With the assistance of a dedicated group of educators and scientists from the University of the South Pacific as well as specialists from various government and non-governmental entities, the UNCA became Fiji's first officially designated Ramsar site on 11 April 2006 through the Convention on Wetlands. Since its inception, the UNCA has yielded two newly discovered endemic species of freshwater fish, sightings of the globally endangered pink-billed parrot finch, and a banded iguana now considered rare in Fiji's forests. In addition to the unique animal life, a surprisingly healthy population of *Metroxylon vitiense*, or sago palm, has been noted. The sago palm has been significantly reduced or eliminated on many of Fiji's freshwater river drainages (Macedru, 2013).

The Convention on Wetlands is 'the first of the modern global intergovernmental treaties on conservation and wise use of natural resources' (Young, 2015, p. 1). The national and international significance of protecting fragile wetlands is critical for the preservation of biodiversity and subsequently the health and well-being of humanity. The UNCA is now part of more than 1280 wetlands around the world designated as areas of international importance that make up nearly 2 million km^2 of surface area, an area larger than France, Germany and Switzerland combined. The Ramsar Bureau in Fiji is housed in the headquarters of the IUCN in Suva, and works in collaboration with other inter-governmental and non-governmental entities that contribute expertise and resources to 'the development of policies, technical and scientific tools of the Convention and to their application' (Young, 2015, p. 1). To date there are four primary international organizations that are associated with the Ramsar, Convention on Wetlands: BirdLife International, The World Conservation Union (IUCN), Wetlands International and the World Wide Fund for Nature (WWF).

A river of ecological significance

The Navua River represents Fiji's third largest freshwater drainage with only the Rewa and Sigatoka rivers draining larger areas. The UNCA, which is located along the river, has approximately 50 waterfalls in the dry season and easily 100+ during the wettest times of the year, approximately 53 mapped major side streams that feed into it, and upper reaches that are channelled into a narrow and sheer-walled canyon creating a funnel effect that sees water fluctuations of 30 ft and greater during severe weather. The upper reaches of the river can play a role in flood control for lower lying areas of the catchment.

In addition, the UNCA is host to important flora and fauna. From an ecological perspective alone this feature makes it especially significant. The surrounding rainforest, river and associated biodiversity provide the necessary supplies and food required for daily living, botanical cures, and a wide range of ecosystem services to the entire catchment area, including flood control for lower areas. More specifically, the waters of the UNCA contain breeding populations of at least two endemic freshwater fish species (*Redigibius leveri* and *Schismatogobius chrysonotus*) and, in the forests surrounding the gorge, 17 endemic species of birds.

Due to its relative inaccessibility, the UNCA is in a somewhat undisturbed state, but increased logging in the area and potential mining exploration pose potential threats. The surrounding area is either under logging lease or, in select areas, used for subsistence farming by local villagers. The rich cultural landscapes, which are confined mainly to the lower portions of the Navua River, include shifting agricultural gardens, grazing lands, trails and tracks, fishing grounds, traditional camping or refuge sites, secondary forests and agroforests, and two villages. All of these landscapes offer considerable potential for ecotourism development and include culturally significant biodiversity and associated history that need to be preserved.

The Navua River constitutes part of the traditional fishing grounds or i qoli qoli for landowners. Countless locations known to be good fishing sites or 'holes' are located along the river. Many have traditional names and are known only to the best fishermen and women. Different types of sites (e.g. deep holes, rapids, tributaries) are favoured habitats for different food species, such as freshwater eels (duna), prawns (ura), crabs (sarakali), shellfish (sici) or the wide range of freshwater finfish (ika) found in the river system. There are also a number of traditional camping sites, caves or safe havens, most of which have names that have been used for centuries by locals (including warriors) and travellers for temporary stays or refuge.

2.3.2 Educational outreach and the UNCA

Despite the overall success of the UNCA, the area continually faces unnecessary stress(ors) on its cultural and geographical resources. Illegal logging within the boundaries of the UNCA have generated a need for protection of the area and for conservation management. Logging threatens the trees and plants of the UNCA as well as the entire ecosystem due to erosion and pollution. Logging also poses a long-term danger to the vitality of the UNCA and its people who live and depend on natural resources for their well-being and livelihoods. In response, RF introduced a sustainable alternative to logging – whitewater tourism – which can support present and future generations of mataqali. It also developed programmes to educate residents about the importance of supporting ecotourism as an alternative strategy for the overall well-being of the area.

According to Wight (1994, p. 40), ecotourism 'should promote understanding and involve partnerships between many players, which could involve government, non-governmental organizations, industry, scientists, and locals'. Thus, through the support of a Ramsar Small Grants Fund, RF developed environmental education programmes for residents living in the UNCA. The goal of the programmes are to educate and inform school children and their parents about the ecological importance, cultural significance and economic viability of the UNCA. The programmes were created in cooperation with individuals and Nature Fiji–Mareqeti Viti (NFMV, a local NGO), various entities within the Fijian government, nine mataqali, and residents of Nabukelevu and Wainadiro, two villages located along the river. NFMV worked closely with RF to deliver the educational component of the outreach programmes. Various stopping points along the river corridor were turned into classrooms where participants learned about the flora and fauna of the conservation area. These riverside classrooms also allowed guides and chaperones to share their traditions and stories about the UNCA that they had been taught by their families and friends. During the first outreach programme, for example, guests were instructed to take pictures of things along the river that struck them as beautiful or important. The pictures were then incorporated into presentations delivered to the communities and villages. The outreach programme impacted guests in numerous ways, as shared through the following quotes by members of a student group:

> I really like the trip since it was my first time up there. I have never known the important role played by the environment in our lives and as landowners we should just take stock and protect it.

> The trip was an educational one. I like the voices of nature when the birds were singing and seeing how flowing water build rivers. It was an amazing sight. No wonder the tourists always like to visit the area as it brings us closer to the earth as our source of life. I believe we should protect it for our future and the coming generation.

> I like the idea of going up there, because if it was not for Mareqeti Viti and Rivers Fiji, I believe we would be blinded from seeing and appreciate our natural environment. I urge you all listening today not to think of ourselves only but to think about tomorrow and our children as well.

2.3.3 Direct benefits and contributions from tourism

In addition to conservation, the RF ecotourism project provides direct socio-economic benefits through its tourism operation including, but not limited to, direct employment, philanthropic efforts and community initiatives.

With respect to direct employment, RF set up several strategic processes to facilitate employment of mataqalis in day-to-day operations as well as long-term sustainability initiatives. The company employs 24+ staff from local communities. From the general manager to the equipment truck drivers and mechanics, all employees are local residents from villages along the river and within 20 miles of RF base of operations.

RF's philanthropic efforts are best exemplified through the following quote (personal communication, G. Wendt, 15 May 2012):

> It is our hope through participation in Rivers Fiji programs, guests leave with a greater appreciation for the role tourists can play in conservation of culture and environment. Rivers Fiji believes this heightened awareness to the fragility of our planet is utilized in the critique and evaluation of other activities our guests choose in the future. We know that in part this is being achieved when participants have written back to offer assistance in conservation in Fiji.

Contributions (e.g. money, goods, services, professional assistance) that enhance the quality of life of local communities throughout the UNCA have increased due to the introduction of tourism. They must be managed, however. RF guides encourage guests not to contribute directly to villagers individually to avoid disproportionate distribution of goods/materials and expectations. They ask guests to donate to specific community projects. For example, donations have been used to buy schoolbooks/materials for local schools and community groups, funds have been raised through the sale of rugby shirts and kava at the RF store to support the mataqali, and a partnership has been developed with a medical residency programme from the USA to provide community health support services annually, in conjunction with the provincial health departments.

Community initiatives that facilitate the participation of each Mataqali in RF's day-to-day operations as well as long-term sustainability initiatives include:

- Meetings with and the distribution of surveys to mataqali to obtain feedback on RF operations. The information gathered is used to make key culture- and environment-related decisions (i.e. rules for guests, days of operation, decision-making processes, traditional uses of land, respect for sacred burial grounds).
- Asking mataqali to approve the lease/operational management plan associated with the UNCA, and to utilize traditional systems to create the rotational guide and establish the employee hiring process.
- With the support of mataqali and the Nature/Maraqeti Fiji NGO, developing educational river trips for children and their parents, as a way of gaining long-term support for the UNCA, and river conservation.
- Working with Tourism Fiji to establish conservation workshops and mechanisms for protection.
- Participating in environmental awareness campaigns and the organization of international events/projects (e.g. IMAX film, Eco-Challenge), including the development of 'River of Eden', a film that highlights the UNCA and the partnership for conservation.
- With respect to its guides, promoting continued education by offering a 3-month comprehensive training programme and financial support for attendance at conferences and workshops. RF also encourages guides to: (i) share important aspects of their local customs and traditions as well as their knowledge of the environment with guests; and (ii) assist in the development of interpretive materials and new services.

Given its overall commitment to and respect for Fiji, its indigenous people and the environment, it is not surprising that response to RF UNCA operations has been positive (see Beeftink, 2004; Kerstetter and Bricker, 2012). To date, mataqalis have exhibited their support for RF by thwarting a logging company's plans to build a road through the UNCA and extending the lease programme and fully accepting the UNCA River Management Plan proposed by RF.

2.3.4 Guides as stewards

With local community members as employees, a greater level of conservation stewardship has occurred, especially with the guides. Locally born and raised, RF guides are natural stewards of the land and their culture. They address local logging and mining operations and their impacts. They also share 'Leave No Trace' and minimum impact techniques with guests to help protect the environment and enhance their experience. To help guests understand the indigenous culture of Fiji, guides provide a briefing for guests visiting villages or cultural situations that require particular protocol, including dress and behaviour (e.g. what to do during a kava ceremony). Guides also point out aspects of village life as guests float past rural communities. Their first-hand knowledge of traditions and lifestyle provides guests with a special window on rural Fiji and culture. The experience is not 'staged' and what is seen and experienced is the daily life of the rural highlands Fijian people. In support of its employees' efforts, RF also provides guests with an overview of potentially damaging projects, information on the conservation area, and a comprehensive review of mataqali's land, people and heritage.

2.4 Conclusions

We have provided numerous examples of how RF has partnered with various stakeholders to create, maintain and ensure an extended future for the UNCA. First, through the establishment of the UNCA, the only lease for conservation within Fiji and perhaps the entire South Pacific, a unique partnership was established to provide long-term economic benefits to local communities through an ecotourism product (i.e. whitewater rafting). In turn, the development of the UNCA enticed scientists and the conservation community to explore the UNCA as a candidate for Fiji's first Ramsar site. This process led to educational opportunities for residents, tourists, the government and conservation enthusiasts, which have helped to protect Fiji's third largest watershed and its unique flora and fauna.

Second, the ecotourism lease developed by RF and mataqali of Serua Province has introduced tourism dollars that support conservation of the UNCA. As Fiji's first Ramsar site, the UNCA allows for sustainable use of natural resources by landowning groups and protects the unique natural resources within the river canyon (RF, 1997). As one community member noted (Bricker 2001, p. 242):

> Tourism is very different from other industries such as the logging industry or the fishing industry. The logging and the fishing industry will take away from our land our resources . . . whereas, tourists would not take away any of our resources. They will just come, watch and experience our resources that also include our culture, and us the people. When the logging company departed from here, they departed with our forests. The land is stripped bare. When the fishing company departed from our waters, they departed with our marine life and our waters are scarce or even extinct with marine life. Our resources must be safeguarded so that our future generations can also utilize them.

Third, tourism products created as a result of the UNCA have generated social benefits. They have allowed for the integration of traditional practices into community work projects (Bricker *et al.*, 2012) and increased local control. Local control has helped with: adjustment to tourism development (Beeftink, 2004); establishment of guidelines for visitation (e.g. whether there should be trips on Sundays, what community service projects should be implemented); and guidance for business operations, including employment rotations, employment shifts and management (Bricker, 2001; Beeftink, 2004).

And, finally, Beeftink (2004), Bricker (2001) and Bricker *et al.* (2012) indicated that five overarching aspects of RF's approach were integral to the success of the UNCA project. These five aspects should be of particular interest to ecotourism companies/conservation organizations in island states/countries that are interested in implementing a lease for conservation.

1. Adopting a slow start: RF developed a comprehensive management plan that included hiring and firing guides, days of operation, and traditional uses of the river corridor (Bricker, 2001). Mataqali were consulted throughout the development process, which, while time consuming (nearly 3 years in the making), proved to be beneficial in solidifying relationships and generating the structure for the lease and eventual wetland designation (Bricker, 2001; Bricker *et al.*, 2012).
2. Ensuring an integrated approach: input from NGOs, tour operators, local businesses and residents was crucial to the design of products and distribution of proceeds/benefits. The integrated approach resulted in a signed lease for conservation for 25 + years – the first of its kind in Fiji (Bricker, 2001; Bricker *et al.*, 2012).
3. Education and training: the UNCA project included an educational component targeted to tourists, local residents, the national government and other tourism operators. This has assisted in terrestrial conservation initiatives in Fiji, such as the Sovi Basin with the Fiji National Trust (Bricker, 2001; Bricker *et al.*, 2012).
4. Maximizing local benefits: through local control and direct allocation of resources to conserving natural habitat, the maximization of local benefits has been achieved (Bricker *et al.*, 2012). For example, paying mataqali for the number of tourists that pass through their land has led to increased interest in and care for the lease arrangement as opposed to one-off opportunities for payment (Bricker, 2001; Bricker *et al.*, 2012).
5. Evaluation and feedback: changes in the economic structure, social structure and environment must be monitored closely. If not, destinations or tourism providers can get caught up in trying to meet the demands of tourists rather than focusing on the sustainability of the product. Rivers Fiji conducts regular meetings with mataqali on matters of operations, guide training, education programmes and matters of concern to the landowning groups. These meetings form important touch points on how the communities view RF daily operations, updates on road and logging challenges, as well as provide RF with an opportunity to listen to the needs of communities, learn of any changes in leadership and generally provide an update on business operations. In addition, surveys and photo elicitation techniques have been used to evaluate the community education programmes. Examples of evaluative measures that could be used in the future include: on-going focus group discussions with

visitors, community members, stakeholders and others who have a vested interest in the conservation area; continue photo elicitation studies with residents to document what is special and cannot be jeopardized by tourism in the protected area; storytelling to document the unique traditions linked to the protected areas, and more.

The case study on RF has demonstrated that a small ecotourism tourism company can work with multiple partners to generate a sustainable environmental/conservation management plan that directly benefits those most directly impacted by development. The unique partnership model created by RF brought economic alternatives to rural communities and extraordinary experiences to visitors, while conserving critical wetlands. Equally important, the partnership model is replicable and can be used in other areas deserving of protection.

Notes

1 Mataqali are descendants from common ancestors, traditionally with specialized roles; typical land-owning unit (Walsh, 2006, p. xi).

2 A district is a sub-unit of a province that forms administrative frameworks (Walsh, 2006). Fiji is divided into 17 districts, each with a district officer and five sub-districts with assistant district officers. The districts generally centre on towns and cities, but some follow provincial or tikina boundaries. A tikina is a territorial unit, which is smaller than a province (Walsh, 2006).

3 The new Land Use Decree provides indigenous landowners with an alternative to administer i Taukei lands and challenges the functions of the TLTB as the sole administrator for i Taukei lands.

References

Aalbersberg, B., Tawake, A. and Parras, T. (2005) Village by village: Recovering Fiji's coastal fisheries. In: United Nations Development Programme, United Nations Environment Programme, and the World Bank and World Resources Institute (eds) *World Resources 2005 – The wealth of the poor: Managing ecosystems to fight poverty.* World Resources Institute, Washington, DC, pp. 144–152.

Beeftink, K. (2004) Perceptions of ecotourism: A case study of whitewater guides in the rural highlands of Fiji. MSc thesis, West Virginia University, Morgantown, West Virginia.

Bricker, K.S. (2001) Ecotourism development in the rural highlands of Fiji. In: Harrison, D. (ed.) *Tourism and the Less Developed World: Issues and case studies.* CABI, Wallingford, UK, pp. 235–250.

Bricker, K.S., Kerstetter, D. and Beeftink, K. (2012) Tradition and place: Tourism development from highlands to islands in Fiji. In: Dodd, J. and Sharma, V. (eds) *Leisure and Tourism: Cultural paradigms.* Rawat Publications, Jaipur, India, pp. 66–95.

Clarke, P. and Gillespie, T. (2008) *Legal Mechanisms for the Establishment and Management of Terrestrial Protected Areas in Fiji.* IUCN Regional Office for Oceania, Suva, Fiji.

Clarke, P. and Jupiter, S. (2010) Law, custom and community-based natural resource management in Kubulau District (Fiji). *Environmental Conservation* 37(1), 98–106.

Fiji Islands Bureau of Statistics (2015) Fiji Islands Bureau of Statistics, Fiji facts and figures. Available at: http://www.statsfiji.gov.fj/statistics/population-censuses-and-surveys (accessed 28 June 2016).

Govan, H., Tawake, A., Tabunakawai, K., Jenkins, A., Lasgorceix, A., Schwarz, A.M., *et al.* (2009) *Status and Potential of Locally-managed Marine Areas in the South Pacific: Meeting nature conservation and sustainable livelihood targets through wide-spread implementation of LMMAs.* SPREP/WWF/WorldFish-Reefbase/CRISP, Suva, Fiji.

Govan, H., Jupiter, S. and Comley, J. (2012) Recognition and support of ICCAs in Fiji. In: Kothari, A., Corrigan, C., Jonas, H., Neumann, A. and Shrumm, H. (eds) *Recognising and Supporting Territories and Areas Conserved By Indigenous Peoples and Local Communities: Global overview and national case studies.* Technical Series No. 64: Secretariat of the Convention on Biological Diversity, ICCA Consortium, Kalpavriksh, and Natural Justice, Montreal, Canada, pp. 1–32.

Harrison, D. (1999) *Ecotourism and Village-Based Tourism: A policy and strategy for Fiji.* Ministry of Tourism and Transport, Suva, Fiji.

International Union for the Conservation of Nature (2008) Definition of protected area. Available at: https://www.iucn.org/about/work/programmes/gpap_home/pas_gpap (accessed 15 July 2015).

Jenkins, A.P. and Jupiter, S.D. (2011) Spatial and seasonal patterns in freshwater ichthyofaunal communities of a tropical high island in Fiji. *Environmental Biology of Fishes* 91, 261–274.

Johannes, R. (1978) Traditional marine conservation methods in Oceania and their demise. *Annual Review of Ecology and Systematics* 9(November), 349–364.

Jupiter, S., McClennen, C. and Matthews, E. (2012) Vatu-i-Ra Seascape, Fiji. In: Hilty, J., Chester, C. and Cross, M. (eds) *Climate and Conservation: Landscape and seascape science, planning and action.* Island Press, Washington, DC, pp. 148–161.

Kerstetter, D. and Bricker, K.S. (2009) Exploring Fijian's sense of place after exposure to tourism development. *Journal of Sustainable Tourism* 17(6), 691–708.

Kerstetter, D. and Bricker, K.S. (2012) Tourism and residents' quality of life. In: Uysal, M., Perdue, R. and Sirgy, J. (eds) *Tourism and Quality of Life Research: Enhancing the lives of tourists and residents of host communities.* Springer Publishers, New York, pp. 445–461.

Kerstetter, D., Bricker, K.S. and Li, H. (2010) Vanua and the people of the Fijian Highlands: Understanding sense of place in the context of nature-based tourism development. *Tourism Analysis* 15(1), 31–44.

Locally-Managed Marine Area Network (LMMA) (2008) Coming together: sharing lessons from the 2008 LMMA network-wide meeting community exchange. LMMA, 5–7 November 2008, Fiji. Available at: https://www.issuelab.org/resource/coming_together_sharing_lessons_from_the_2008_lmma_network_wide_meeting_community_exchange_5_7_november_2008_fiji (accessed 15 July 2015).

Macedru, K. (2013) *The Upper Navua Conservation Area: Fiji's first wetland of international importance.* Maraqeti Fiji and Rivers Fiji, Suva, Fiji.

Mills, M., Jupiter, S., Pressey, R., Ban, N. and Comley, J. (2011) Incorporating effectiveness of community-based management strategies in a national marine gap analysis for Fiji. *Conservation Biology* 25, 1155–1164.

Ministry of Fisheries and Forestry (2007) Fiji forestry policy statement. Republic of Fiji, Suva, Fiji. Available at: http://fiji-reddplus.org/sites/default/files/pafpnet_Fiji%20Forest%20Policy%20Statement%20Nov%202007.pdf (accessed 2 August 2015).

Republic of Fiji (2014) *Economic and Fiscal Update: Supplement to the 2015 budget address.* Ministry of Finance, Suva, Fiji.

Rivers Fiji (RF) (1997) *Rivers Fiji Management and Operations Plan.* Pacific Harbour, Republic of Fiji.

Veitayaki, J. (1997) Traditional resource management practices used in the Pacific Islands: An agenda for change. *Ocean and Coastal Management* 37(1), 123–136.

Vukikomoala, K., Jupiter, S., Erasito, E. and Chand, K. (2012) *An Analysis of International Law, National Legislation, and Institutions as they Interrelate with Territories and Areas Conserved by Indigenous People and Local Communities.* Natural Justice in Bangalore and Kalpavriksh, Delhi, India.

Walsh, C. (2006) *Fiji: An encyclopedic atlas.* The University of the South Pacific, Suva, Fiji.

Wight, P. (1994) Environmentally responsible marketing of tourism. In: Cater, E. and Lowman, G. (eds) *Ecotourism: A Sustainable Option?* John Wiley & Sons, Chichester, UK, pp. 39–53.

Wildlife Conservation Society (2009) *Ecosystem-based Management Plan: Kubulau District, Vanua Levu, Fiji.* Wildlife Conservation Society, Suva, Fiji.

World Bank (2015) Country Statistics of Fiji. Available at: http://data.worldbank.org/country/Fiji (accessed 2 August 2015).

Young, L. (2015) Conserving Fiji's Wetlands. The RAMSAR Convention on Wetlands. Available at: http://ramsar.rgis.ch/cda/en/ramsar-news-archives-2012-fijireportunca/main/ramsar/1-26-45-520%5E25939_4000_0__ (accessed 21 August 2015).

3 Rhetoric Versus the Realities of Sustainable Tourism: The Case of Cyprus

Anna Farmaki,[1*] Levent Altinay[2] and Muhammet Yaşarata[3]

[1]*The Cyprus University of Technology, Cyprus;* [2]*Oxford Brookes University, UK;* [3]*Cyprus Premier Holidays, UK*

3.1 Introduction

The belief that sustainability in tourism should be an integral part of tourism policy aimed at balancing the costs and benefits of tourism is widespread and time honoured, exemplified by the innumerable academic papers and books on tourism sustainability and the support by intergovernmental and private tourism institutions. Defined as 'the forms of tourism which meet the needs of tourists, the tourist industry and host communities today without compromising the ability of future generations to meet their own needs' (Swarbrooke, 1999: 13), sustainable tourism has often been presented as the antidote to mass tourism problems. Understandably, sustainability in tourism gained considerable attention as attempts to conceptualize the term and understand its principles were made. Initial conceptualization centred on the environmental aspects of tourism development, with ecotourism and green tourism becoming synonymous to sustainability. Over the years the term 'sustainability' was reshaped indicating a transition to social and economic considerations. Similarly, efforts to redefine the term and its relevance to the tourism industry prompted a debate over the difference between sustainability in tourism and sustainable tourism. Liu (2003) concluded that whilst the sustainable development of tourism should be viewed as a process-oriented, state-focused goal associated with the managed change towards improvement, sustainable tourism refers to all forms of tourism (conventional and alternative) that need to be compatible to sustainable development principles. Hence, the sustainable development of tourism does not necessarily signal a departure from mass tourism. The synthesis of the polyglot of definitions and concepts leads to a general acceptance that sustainable tourism is based on the three pillars of economic, social and environmental sustainability and that a holistic approach in tourism policy should be adopted to encourage social

*E-mail: anna.farmaki@cut.ac.cy

learning and adaptive management. The link of sustainable tourism development to policy making denotes a managerial or political element in its conceptualization, with scholars arguing that effective and inclusive governance is required for the successful implementation of sustainable tourism (Scheyvens, 2011).

Hence, sustainable development is a critical success factor for both governments and the tourism industry. First, sustainability in tourism has been argued to improve destination competitiveness. Ritchie and Crouch's (2003) milestone study on destination competitiveness proposes that what makes a destination competitive is its ability to maximize tourism expenditure, increase the number of visitors and provide a satisfying experience to tourists in a profitable way while ensuring the well-being of the local community and the preservation of the environment. Hence, destination competitiveness is a multi-dimensional concept that encompasses economic, environmental, socio-cultural, technological and political factors. There is notable work arguing the contribution of sustainability in destination competitiveness (Bahar and Kozak, 2007; Mazanec *et al.*, 2007; Crouch, 2010). For instance, empirical research illustrates that an environmentally sustainable tourism industry responds to the growing needs of a 'green market' (Mazanec and Ring, 2011). As Cucculelli and Goffi (2016) report, a sustainable tourism policy is strongly associated with destination competitiveness, to which natural and cultural resources, an authentic experiences and quality service contribute.

Additionally, scholars corroborate the idea that a competitive destination maximizes the socio-economic prosperity of the local community as it leads to an improved standard of living (Dwyer *et al.*, 2004). Therefore, sustainable tourism development improves quality of life of the host community. Although there is no clear definition of quality of life, it is generally understood as one's satisfaction with their life. An improved quality of life transpires through a higher standard of living, higher tax revenues, increased employment opportunities and economic diversity (Andereck *et al.*, 2007). In addition to economic well-being, tourism impact studies have enlisted social, cultural and environmental well-being of a community as additional effects of tourism development. For instance, a person's quality of life might be improved through interaction with a cleaner environment or due to increased pride over their culture. Specifically, Andereck and Nyaupane (2011) suggested that factors such as peace, safety, clean air and water and infrastructural services play a role in residents' perceived quality of life. Indeed, through sustainability several of the negative impacts of tourism can be alleviated (i.e. crowding, degradation of the environment and loss of traditional culture). Moreover, sustainable tourism development empowers local communities as it promotes inclusive decision-making and planning. Kim *et al.* (2013) argued that residents' perception of quality of life varies according to the stage of tourism development cycle. As destinations become more mature the negative impacts of tourism development increase. Hence, rejuvenating the destination through sustainable tourism forms can lead to the enhancement of community perceptions of quality of life.

Sustainable tourism has also been argued to extend the seasonal nature of destinations' tourism industries. Martin *et al.* (2014) highlighted the adverse relationship between seasonality and sustainability, as seasonal tourism trends threaten sustainability and impact local communities socially and economically. For instance, seasonal tourism limits investors' ability to yield returns on their capital and burdens them with fixed costs off-season. Similarly, tourists pay higher prices for goods during the peak season whereas tourism employees are faced with long periods of unemployment (Cuccia and Rizzo, 2011). In addition, the congestion experienced during high season tends to worsen the relationship between hosts and tourists. As Nadal *et al.* (2004) commented, seasonal patterns of tourism activity can differ depending on the destination lifecycle stage, yet the outcome of seasonality seems to be exacerbated at the maturity stage of many island destinations which rely on coastal tourism. Thus, sustainable tourism development through the development of special interest activities, such as agrotourism and cultural tourism, and the targeting of niche tourist segments can significantly improve the costs inflicted by seasonal tourism demands.

The importance of sustainability in tourism becomes particularly germane to island tourism contexts. *A fortiori*, island destinations are characterized by the existence of fragile eco-environments, limited water resources, rival land uses and inadequate waste management infrastructure (Baldacchino and Ferreira, 2013). In particular, the synonymous development of island tourism with the mass tourism phenomenon exacerbated the vulnerability of islands. Tourism developments in islands in the Mediterranean basin are characterized by a reliance on the commercialization of islands' coastal regions, a highly seasonal tourism product, which is based on the 3S (sea, sand, sun) components, and a dependency on foreign tour operators and low-cost carriers that transport 'golden hordes' of mass tourists to islands. Additional problems of tourism development in islands have been reported including conflict between the host community and tourists, environmental degradation, abandonment of inland areas as islanders opt for employment opportunities in coastal regions and seasonal unemployment among others (Andriotis, 2004). Therefore, a sustainable approach to tourism development within island contexts has been widely regarded as an incumbent strategy that has been long overdue (Ko, 2005).

None the less, empirical evidence illustrates that in most instances island destinations are challenged in their attempt to transfer the rhetoric of sustainability into reality (Bianchi, 2004; Altinay and Hussain, 2005; Dodds and Kelman, 2008; Yaşarata *et al.*, 2010; Farmaki *et al.*, 2015). Factors identified as inhibiting can be grouped into several categories: (i) factors pertaining to the limited understanding of the concept of sustainability and a general lack of sustainability awareness; (ii) factors relating to the lack of governmental support in terms of the sustainable development of tourism; (iii) factors stemming from the poor coordination of activities between national and local stakeholders, as well as between the public and private sectors; (iv) a weak decision-making structure that marginalizes certain interested groups from policy design; and (v) environmental constraints and limited employment opportunities that islands offer. This chapter aims to offer insights into the paradigm of sustainable tourism by drawing on empirical evidence from the island of Cyprus in relation to the five categories of factors mentioned above. Discussion centres on the interplay of destination competitiveness, seasonality and quality of life.

The chapter begins by presenting the theoretical background necessary in framing discussion into the relevant context. Further on, a synopsis of empirical research on the sustainability of island tourism is provided before the chapter introduces the case of Cyprus. Conclusions from selected examples of sustainable tourism practices on Cyprus are given and the chapter ends with recommendations on how sustainable tourism in Cyprus can be better implemented to serve the goals of improved destination competitiveness, seasonality and quality of life.

3.2 Sustainable Island Tourism: From Conceptualization to Implementation

The increase in the number of tourists visiting small island destinations from 28 million in 2000 to 41 million in 2013 indicates the popularity of islands as destinations (UNWTO, 2014). Apostolopoulos and Gayle (2002) attribute the growth of island tourism to transportation improvements in the post-Second World War era, coupled with the growth of mass tourism in the late 1960s, which encouraged the robust development of island tourism in the Mediterranean, Caribbean and Pacific regions. As island communities recognized the potential of tourism in generating economic prosperity through foreign exchange, tax revenues, economic diversification and employment opportunities, they popularized tourism as a developmental option. None the less, the geographic, cultural, ecological and economic characteristics that make islands attractive create a fragile context forcing island communities to succumb to the pressures of tourism development. Among the weaknesses of islands are a fragile ecosystem, limited water and land resources, infrastructural deficiencies in transportation and communication as well as economic and political adversity such as heavy dependence on imported goods and foreign labour, economic leakage and political subordination to former colonial forces. The vulnerable context of islands is further problematized by reliance on mass tourism, which has been criticized as a failed approach to tourism development.

Scholarly echoes emphasizing the need for islands to adopt sustainable development practices are not unjustified and largely shared by the United Nations (UN) and other international non-governmental organizations (NGOs). Twining-Ward and Butler (2002) warned that sustainability in islands is a necessity rather than an optional extra. None the less, the transition towards sustainability has been challenged by two factors: (i) the semantic confusion surrounding the conceptualization of the term sustainability; and (ii) the difficulty in operationalizing and applying the paradigm in island contexts. The outpour of literature seeking to clarify the concept of sustainability is evidence to the lack of consensus over its nature, objectives and feasibility. For island tourism, sustainability is understood as an approach aiming at 'preserving a steady stream of income by creating an adaptive competitive destination niche market through the continuous guidance of participatory community planning without sacrificing the socio-cultural and natural integrity of the asset base' (McElroy and Albuquerque, 2002: 16). Hence, it can be argued that the primary goals of sustainable island tourism centre on the improvement of the quality of life for the island community, the preservation of the environment and traditional culture, inclusive and participatory tourism planning and the improvement of the destination's competitiveness through the provision of an enhanced quality of experience for visitors.

Nevertheless, definitional attempts of sustainable island tourism are ambiguous for several reasons. First, it is not clear which type of activities or strategies are needed to develop sustainable tourism. As Lim and Cooper (2009) stated, activities such as carrying capacity management, local community involvement and the development of special interest tourism might be the stepping stone for islands to aspire to the goals of sustainability. Yet, such activities may not always yield the desired effect. For instance, culture- and nature-based niche products have been highly promoted as improving destination competitiveness by extending seasonality, increasing tourist spend and targeting additional segments of visitors. On the other hand, the development of such products acquires additional investment on natural and cultural resources, which inflicts further pressures on the environment. Also, the economic dimension of the sustainability approach places the issue of carrying capacity under scrutiny. Second, while the empowerment of the local community is a key target of sustainability, conceptualization attempts fail to acknowledge the influence of residents' perceptions and attitudes towards tourism development as well as their ability to develop tourism sustainably. Last, the evaluation of sustainability in tourism, founded on the vague roots of the term's conceptualization, is open to interpretation and misuse. As Choi and Sirakaya (2006) noted, the lack of precise criteria of sustainability in island tourism inhibits its implementation. As a response, scholars have directed their efforts at establishing indicators for sustainability in tourism, highlighting their essentiality for: (i) monitoring tourism development and performance through the assessment of related policies and practices; and (ii) communicating appropriate knowledge through quantifiable data (Roberts and Tribe, 2008). None the less, while economic and environmental aspects of sustainability can be relatively easily measured, the ambiguity surrounding the subjective interpretation of the social dimension of sustainability limits its quantifiable scope.

In general, indicators of sustainable tourism can be divided into several categories according to the goals of the paradigm: (i) indicators pertaining to the environmental and ecological preservation of destinations; (ii) indicators related to the economic performance of tourism activity and the sectoral contribution to the economy overall; (iii) indicators associated with societal well-being and local community satisfaction with their life; and (iv) institutional sustainability indicators, namely those factors that illustrate effective and inclusive governance and decision-making structures (Table 3.1).

Despite the notable attempt to identify appropriate sets of indicators for sustainable tourism, there is criticism over their narrow focus and tourism-centric orientation. In addition, critics questioned the transferability of the indicators, arguing that sustainable tourism development is time and context specific and a universal framework developed to fit all cases is insufficient. Indeed, the plethora of empirical evidence on tourism impacts on islands illustrates varying degrees of success. For example, in Mallorca alternative conservation and rehabilitation of the local community was achieved through effective and inclusive planning (Loukissas and Triantafyllopoulos, 2002). Similarly, Malta was successful in rebranding itself as a cultural destination and targeting an upmarket tourist segment (Chapman and Speake, 2011). Yaw (2005) reports the Caribbean's success in sustaining its tourism industry and responding to the environmental awareness of tourists through cleaner technologies. On the other hand, in Crete the construction of illegal buildings and political pressures by formal and informal groups discouraging environmental protection plans are evidence of a flawed implementation of sustainable tourism (Briassoulis, 2003). Thus, the successful development of sustainable island tourism is inherent to the contextual needs of islands and is subjected to exogenous influencing factors. For instance, an optimum capacity level might bring environmental benefits yet it requires the sacrifice of economic benefits, a tactic which is often not easily accepted by island communities exposed to limited employment opportunities.

Table 3.1. Indicators of sustainable island tourism.

Environmental	Ecosystem protection
	Protection of coastal and marine resources
	Energy/water-saving management
	Conservation of land and forest resources
	Minimize waste and pollution (i.e. recycling)
Economic	Minimize economic leakage
	Reduce seasonal employment
	Maximize economic benefits (i.e. tourism revenues)
	Diversify economic development through the development of niche products (i.e. rural tourism development)
	Increase destination competitiveness by improving tourist experience and enhancing quality of serving and facilities
Socio-cultural	Maximize social benefits from tourism
	Increase perceived quality of life
	Community empowerment (i.e. promotion of local culture and participation in tourism planning)
	Training and educating employees (i.e. rejuvenate traditional arts/crafts)
Institutional	Align tourism planning with general sustainable development policy
	Promote inclusive governance and decision-making
	Coordinate related activities
	Create awareness on sustainability

Likewise, the pervasive nature of the interconnectivity of the small-scale economic, political and social spheres of islands pose challenges to management mechanisms.

Considering the vulnerable environment of island destinations, it is imperative that empirical research sheds light on this imminent topic by acknowledging the factors hampering sustainable island tourism and proposing strategies that can strengthen sustainability attempts. Despite islands' economic and natural constraints, there is optimism among scholars that the necessity for sustainability can provide the motivation for remedial action. It has been argued that sustainability in island destinations can enhance self-sufficiency, extend seasonality, enhance quality of life for local populations, upgrading the environment and improving the competitiveness of island destinations that are in decline (Kokkranikal *et al.*, 2003). Against this background, the following sections of the chapter evaluate sustainable tourism development on the island of Cyprus in terms of economic, environmental and social dimensions. The island of Cyprus, being divided since 1974 following inter-communal conflict, presents an interesting case due to the different stages of tourism development noted in the respective communities of Greek and Turkish Cypriots. Thus, two examples of sustainable projects implemented on the island by each community are discussed and used to serve as paradigmatic cases outlining the nature of sustainable tourism implementation and the degree of implementation success. Challenges are identified before recommendations are offered.

3.3 Tourism Development in Cyprus: a Sustainable Perspective

Cyprus, the third largest island in the Mediterranean, gained its independence from the British in 1960. With an agrarian economy, the newly-founded government of Cyprus recognized the benefits of tourism development and transformed its economy to a tourism-based one. Despite a steady increase of arrivals to the island, inter-communal tensions between the Greek and Turkish Cypriot inhabitants of the island brought its tourism to a halt following the 1974 war. Since then, the island remains divided with the southern part being administered by the Republic of Cyprus (RoC; an internationally recognized state and a member of the EU) and the northern part being administered by the Turkish-Cypriot community that remains a non-recognized state, economically and politically dependent on Turkey.

Interestingly, tourism development in each community reflects a different story. Primarily appealing to northern European markets, tourism development in the RoC coincided with the mass coastal tourism explosion in the 1970s. Over the years, RoC saw its tourist arrivals increase significantly and by the end of the 1990s, 2.5 million tourists were arriving annually on the island yielding almost €1927.7 million in tourism revenue (CTO, 2001). The rapid growth of tourism resulted in several negative outcomes including environmental degradation, unskilled foreign labour, perishing cultural identity and a persistent sea and sun image. By the 2000s, it was clear that the tourism market had reached stagnation and was being further threatened by emerging competition and changing tourist needs. With tourist arrivals fluctuating throughout the last decade, the tourism authorities highlighted the need to adopt a more sustainable developmental approach in tourism and initiated a repositioning strategy by focusing primarily on targeted marketing, the distribution of benefits to local communities and new product development, with a primary focus on culture and nature. The development of sustainable tourism also aimed at de-seasonalizing the main 'sea and sun' product, minimizing environmental pressures on the coastline and preserving traditional culture.

On the other hand, the Turkish-Cypriot community struggled to establish itself in the international tourism arena due to its non-recognized status. Forced into acute financial and political dependency on Turkey (Alipour and Kilic, 2005), the northern part of the island has a small internal market and is unable to attract foreign investment due to political and economic isolation. The attempt to promote tourism is further impeded by limited accessibility, with international tourism arrivals barely reaching 257,000 and tourism revenue amounting to US$459.4 million in 2012, as opposed to the 2.4 million international arrivals and €1927.7 million achieved in RoC (Farmaki *et al.*, 2015). Hence, tourism in the northern part of the island remains highly dependent on the Turkish market with a large majority visiting the island for gambling purposes. In recent years, concerns have been expressed primarily by private-sector actors with unplanned development, dependency on foreign stakeholders and lack of clear policies on tourism being among the issues of concern (Altinay and Hussain, 2005). Therefore, in contradiction to RoC, which pursues sustainable tourism as a response to market volatility, increased environmental consciousness and the negative effects of mass tourism, sustainable tourism in the northern part of Cyprus is driven by private and non-governmental stakeholders in an attempt to capitalize on perceived environmental quality gains. The following section of the chapter will outline the challenges of sustainable tourism implementation by drawing examples from two cases: the Alagadi Special Protected Area Environmental Plan (ASPAEP) in the northern part of the island and the optimization of the All-inclusive Optimisation Programme, initiated by the Cyprus Sustainable Tourism Initiative (CSTI) in the area of Pafos (RoC).

3.3.1 Alagadi Special Protected Area Environmental Plan

Alagadi is well-known for its sandy beach, beautiful natural scenery and surrounding environment. Since 1992, when researchers from Glasgow University were invited by environmental groups in northern Cyprus to monitor marine turtle nesting activities, Alagadi's popularity extended as it was established as a reproduction area in the Mediterranean Sea for the loggerhead turtle (*Caretta caretta*) and the chelonian sea turtles. Hence, the area represents one of the most important environmental resources on the island. Consequently, environmental groups such as the Society for the Protection of Turtles, the Environmental Conservation Department, the Department of Town Planning and the Tourism Planning Department have worked together to develop a plan that placed the area under protection. The aims of the development plan, titled 'Alagadi Special Protected Area Environment' (ASPAEP), were twofold: (i) to protect the turtles' nesting areas and surrounding region; and (ii) to serve as a benchmark offering the necessary knowledge and guidance with regard to planning control and activities.

A starting point for the implementation of the ASPAEP was to divide the 'Turtle Protected Area' into three zones: primary and secondary protection areas and a buffer zone. The primary protection area was where the marine turtles laid their eggs and thus development or activity that endangered the turtles was prohibited. In the secondary protection area and in the buffer zone the ASPAEP aimed at developing the landscape around buildings to conform the area's natural landscape and scenery. Hence, in this zone the construction–plot ratio near the coastline was kept low at a 0.005/1 scale whereas the height of buildings was restricted to only one floor. The development of related facilities and services such as cafeterias, toilets and beach guard towers were also kept to a limited number. Additionally, the plan made provisions regarding the surrounding areas known as the 'settlement areas', where housing developments would take place. The ASPAEP outlined the type of housing and tourism facilities that could be constructed, restricting the number of floors of buildings to two floors maximum. Equally, specifications were made for the 'pastoral areas' in the region that were used for agricultural purposes. More specifically, the plan identified between activities that were permitted in the area such as agriculture, forestry, gardening and horse breeding and those that were prohibited, including stock breeding.

The ASPAEP was implemented in 1999, however, the road to implementation was not without obstacles. In fact, implementation of the plan was delayed several times due to the many disputes among local people, especially those who had property in the protected area and who were not content with the construction restrictions outlined in the plan. Fierce arguments arose at community meetings, which often political participants chose not to attend as a result of the pressures from local people. Amidst fears of not being elected again, politicians faced a conflict with the local community and decision-making bureaucrats. On one hand, government planners wanted to proceed with the approval and implementation of the ASPAEP in order to achieve perceived environmental gains and, on the other, the local community aimed at delaying the approval of the plan in order to gain time and influence politicians in preventing the plan's implementation. Thus, community participation at the initial stages of the ASPAEP was a hindrance rather than an advantage. An important point of concern for the local community was that the plan was restricting tourism development in the area. In the eyes of the locals, tourism was the panacea to unemployment and the subsequent emigration of younger generations, and the bearer of economic miracles that could alleviate the economic problems they faced. Ultimately, the ASPAEP was approved after state officials presented to the local community the related environmental and economic benefits. Political desire and support might have led to the implementation of the plan, however, its success was hindered by an unsatisfied local community which saw potential tourism investors abandoning the area after the plan was implemented. As a result, antipathy towards environmental groups and researchers from Glasgow University grew. The local community's discontented attitude was the product of the perception that the government placed too much emphasis on environmental rather than economic problems in the area.

The next section outlines a differing case of sustainable tourism development. The All-inclusive Optimization Programme, initiated in RoC by the CSTI with the support of the Cyprus Tourism Organisation (CTO), exemplifies the attempt of a mature tourism market to counteract the negative effects of mass tourism.

3.3.2 All-inclusive Optimization Programme

The CSTI, affiliated with the UK Travel Foundation, was established in 2006 as a non-profit organization with the goal to promote the development of a sustainable approach to tourism through the preservation, conservation and the protection of the environment, the sensible use of natural resources and the improvement of the social and economic conditions of marginalized rural communities of Cyprus. Since then, the CSTI has implemented various projects promoting the sustainable development of the tourism industry including water and energy saving practices in the accommodation sector, the reduction of plastic consumption and solid waste in participating hotels, the formulation of guidelines on how to make hotels greener, the publication of a handbook outlining village routes as an alternative tourism experience and the 'Greening Cyprus Beaches' campaign (Farmaki *et al.*, 2014). In 2013, the CSTI initiated a study on the optimization of the all-inclusive (AI) tourist spend in the Pafos region, which served as a pilot study area. Representing one of the main holiday choices for tourists in coastal Mediterranean destinations, AI holidays have been accused of discouraging interaction with locals, the development of 'ghost towns' where local businesses and restaurants remain empty and little contribution to the local economy. Consequently, AI holidays reinforce seasonality of a destination as well as the potential of money leakages. In RoC, tourists from the UK arriving to Cyprus on the AI package account up to 40% whereas a staggering 80% of Russian arrivals select all-inclusive holidays (CTO, 2014). Equally, an increasing number of hotels are being renovated to respond to the requirements and standards of AI holidays. It is thus not surprising that the CSTI, with the support of the CTO and the regional tourism authorities and municipality, undertook in collaboration with Brighton University the above-mentioned programme aiming at improving the long-term sustainability of Pafos, increasing tourist spending in the local economy and enhancing tourist experience and future revisit intention.

Past initiatives that have been carried out as a response to the AI impacts include the 'dine around' meal voucher exchange system, hotel market stalls and locally produced merchandizing in retail outlets, tradition-centred tours, the organization of local food and wine festivals and discounts for local establishments. None the less, the success of the above-mentioned schemes has been limited. For instance, the meal voucher exchange system has not proven effective as the owners of AI resorts tend to include restaurants in which they or family members are involved, thus excluding local businessmen that have no affiliation to the hotels. The pretext for this response is that it is easier to manage pricing/charging and that the coordination of a joint marketing approach is unclear. Additionally, local produce had poor visibility in hotels and retail outlets, often being overshadowed by international tourist merchandise whereas excursions promoted in AI resorts focused primarily on general tours on the island, water sports and boat trips rather the rural attractions. Several factors contribute to the limited success of the above-mentioned tactics. These include the lack of identity of the region of Pafos, which is further fuelled by the large number of foreign workers employed in the industry following the growth of mass tourism, and a confusion among tourists over what is Greek and what is Cypriot. The organization of Greek nights and the serving of Greek-style food in AI resorts contributes further to the fuzziness surrounding the lack of a distinct local identity. Another challenge is the limited evidence of collaboration among stakeholders, which is seen as pivotal in the development of sustainable tourism. Trust between ethically-oriented tour operators and local stakeholders has been lost due to previous negative collaborations and misbehaviour, whereas often competitive tendencies among local stakeholders – primarily between hoteliers and local businessmen – lead to a dividing attitude of 'us' and 'them' rather than embracing a win-win collaborative approach. For example, hoteliers are often accused of having economically-driven interests and hence promoting the AI package, whereas hotel owners defend their actions as being the outcome of economic crisis and pressures from foreign tour operators.

As a consequence, to counteract such obstacles the All-inclusive Optimization Programme recommended the organization of festivals in the area to serve as an attraction encouraging AI customers to exit the hotel surroundings. Despite the support from local stakeholders, fears over the interest AI tourists would show on culturally targeted festivals were expressed. Moreover, the programme suggested the creation of an ambassador scheme known as 'ask me I am local', whereby hotel staff would provide information to AI customers over attractions and facilities to visit. None the less, concerns were expressed from local suppliers with regard to the maintenance of control over the suggestions made by hotel staff. Other local stakeholders, including tour representatives, warned against the conflict of interest that might exist between hoteliers and local businessmen and cautioned over the meaning 'local', in an attempt to emphasize the multicultural base of hotel employees in the Pafos region. Quality labels were also suggested by the programme to indicate tourist establishments offering quality service within the principles of sustainability. However, the feasibility of this tactic was regarded as problematic in terms of which organization or authority would be appropriate in awarding these labels. Suppliers of the tourism industry also expressed concerns regarding the applicability of quality labels highlighting the limited use of local products in the industry due to limited resources. To strengthen local identity based on food and drink, a 'meze trail' was recommended in addition to the meal voucher exchange scheme. Nevertheless, both were regarded difficult to implement due to poor restaurant infrastructure and logistical challenges.

The following section discusses and parallels some of the issues identified from both cases in relation to quality of life, seasonality and competitiveness factors. Subsequently, recommendations are given to serve as guidance for future attempts of developing sustainable island tourism.

3.4 Challenges of Sustainable Tourism Development in Cyprus

3.4.1 Lack of sustainability awareness

Within the literature, raising the level of awareness of the local community with regard to the importance of sustainable development in tourism has been regarded as an important element towards the successful implementation of the paradigm. Hitherto, evidence has taught us that lack of understanding of the concept of sustainability can inhibit the implementation process (Dodds and Butler, 2009). In the case of Alagadi, inclusion of locals at the decision-making stage has not proved uneventful, as a misunderstanding of the perceived benefits of environmental conservation and a restricted carrying capacity led to opposition by the local community, which regarded environmental conservation as a strategy hindering economic development. As Miller *et al.* (2010) posited, the provision of information does not necessarily increase awareness levels among the public. Indeed, politicians' attempts to explain the environmental, social and cultural gains of the ASPAEP to locals were not sufficient in changing residents' perceptions. Rather, situational factors and the complexity of assessing practical alternatives can lead to frustrated behaviour towards sustainable tourism development. As the Alagadi case reveals, contrary attitudes of the local community in terms of tourism investment and environmental conservation exist. Such attitudes are the product of not only a poor understanding of the sustainability paradigm but also contextual factors that push the importance of sustainability down the priority list. For example, in RoC, despite sustainability being embedded in tourism plans, the financial crisis experienced in 2013 led to the government redirecting its policy towards a large-scale developmental approach (Farmaki *et al.*, 2015). Ko's (2005) argument that local communities in small islands are often short-sighted in terms of planning and the need for conservation is therefore not unjustified.

3.4.2 Lack of governmental support

The emphasis placed on sustainability in Cyprus transpires in tourism policies, and as the case of Alagadi demonstrates, political will and support can lead to the promotion of sustainability plans. None the less, the inclusion of sustainability in public policy is neither a sufficient tool for developing a sustainable tourism industry nor a long-term sustainable solution. On one hand, several requirements of sustainability in the implementation process remain out of the scope of the government. Whilst governments can provide support through financial incentives and training, evidence from the All-inclusive Optimization Programme in Pafos illustrates the practical difficulty in addressing sustainability issues due to the corporate nature of the tourism industry, which is largely private-sector driven. Indeed, political support of economic development compromises the goals of sustainability. In addition, the absence of infrastructural provisions, which were reported in Pafos, limited the success of the All-inclusive Optimisation Programme. On the other hand, the long-term orientation of the sustainability paradigm entails the commitment and cooperation of both the public and the private sectors as well as the local community. The Alagadi case exemplifies how private interests challenge public orthodoxies for societal well-being and hence comes as no surprise that in small island contexts political consensus is largely directed by voters. Farmaki *et al.*'s (2015) research on the relationship between sustainable tourism and politics in Cyprus supported Scheyvens' (2011) argument that governments often hide behind the rhetoric of sustainability. Specifically, the authors characterized the tourism industry of the island as 'an impenetrable jungle of interpersonal relationships between politicians and powerful business elites, who govern the industry' (Farmaki *et al.*, 2015: 188).

3.4.3 Poor coordination of activities

The alignment of tourism with the sustainable development of other sectors in an island's economy is well documented as an important facet of overall sustainable development (Dodds, 2007). Consequently, cooperation and coordination among sectors and across levels is required if sustainable tourism is to be successfully implemented. None the less, the complicated multi-faceted nature of the tourism industry, which spans across public and private sectors as well as external stakeholders, hinders policy implementation (Dredge, 2006). The network of actors is further expanded as sustainable tourism policies are often made in non-tourism domains. Indeed, as evidenced in the case of Cyprus, the multiplicity of stakeholders involved in decision-making, each with differing interests, inhibits sustainable tourism implementation. It is not surprising for example that in Pafos the meal voucher exchange system was not utilized, due to the wide variety of stakeholders involved in the provision of this service. Sustainability advocates a holistic approach towards development and the integration of interests and tactics. However, such a directive is often incongruous to the antagonistic behaviour of stakeholders within the tourism industry. Different actors in the public–private network strive for different goals as some are interested in ecological sustainability, others for social or economic sustainability with some members' interests being based on organizational development. Indeed, the Alagadi case presents an interesting example, depicting the diversity in interests and activities of various actors in the tourism industry.

3.4.4 Ineffective governance

Effective governance based on inclusive decision-making is a requirement for sustainable tourism implementation (Hall, 2007). The network structure of the tourism industry, however, restricts the effectiveness of tourism governance. As Farmaki (2015) reported, attention should be directed to the socio-cultural, environmental and economic factors shaping policy outcomes as these hinder the effectiveness of governance structures. The cases presented in this chapter offer interesting insights in terms of the factors that restrict the effectiveness of governance in Cyprus. For example, in Alagadi the environmental conservation plan was not easily accepted by the local community due to the economic sacrifices involved as a result of its implementation. As tourism is primarily related to decisions over resource distribution, it becomes apparent that in small island contexts, where resources are limited, competition over resource allocation is fierce. The Alagadi beach and the surrounding area represented an issue of negotiation among stakeholders. Indeed, Farmaki *et al.* (2015) found that power struggles among national and local stakeholders in Cyprus as well as global actors intensify the complexity of the socio-political environment which renders the challenge of sustainable tourism implementation. Another issue of ineffective governance that is well documented in tourism literature is the marginalization of certain groups from tourism planning. The case of the All-inclusive Optimization Programme in Pafos, for example, illustrates how less powerful groups, such as small local businessmen, are not predominantly involved in the development of sustainable tourism plans and thus supports Farmaki *et al.*'s (2015) finding that in Cyprus undemocratic networks prevailing in the tourism industry weaken governance structures, destination management and social learning.

3.4.5 Environmental and economic constraints

Kerr (2005) highlighted that the development profile of islands is restricted by two key factors: (i) issues of scale, which involves limited natural and human resources, diseconomies of scale in infrastructure development and service provision and the monopolistic nature of island economy; and (ii) issues of isolation, which include a high cost of transport, expensive manufacturing, vulnerability to environmental impacts and difficulty in meeting demand in the modern supply chain. Indeed, islands exhibit dependency on foreign forces due to their environmental and economic constraints. Limited in their ability to support their economy through other means, island tourism consequently dominates the political, economic and social spheres of the local community as evidenced in the case of Cyprus. Islands' peripherality and geographic remoteness create a complex economic and political context that relies on exogenous factors for prosperity and development and in which the importance of environmental conservation is minimized. Opposition from locals in Alagadi, for instance, illustrates the concerns of the local community over their economic prosperity if the environmental conservation plan was to be implemented. Evidently, islands become victims of their own fragile environment which offers limited opportunities in terms of employment and growth. Similarly, the reliance on the AI holiday type as exemplified in Pafos demonstrates the dependence of Cyprus on foreign tour operators that promote AI holiday packages. Unsurprisingly, Issa and Jayawardena (2003) commented that mass tourism on islands has become a 'necessary evil', forcing island destinations into a vicious cycle of dependence from which it is difficult to escape.

3.5 Conclusions

Cyprus' economy depends greatly on tourism. Evidently, the sustainable development of the island's tourism industry holds a high place on the planning agenda of politicians. None the less, as this chapter illustrates, sustainable tourism implementation in Cyprus remains a challenging task, entrenched in the problematic context of island destinations. Cyprus is characterized by dependence on foreign tour operators, limited environmental and economic resources while the large number of stakeholders involved in the tourism industry leads to conflicting interests and poor coordination of activities. The island's reliance on tourism gives rise to an economic orientation that becomes prevalent in decision-making and planning as well as local community attitudes towards sustainability. Whilst the promotion of the environmental conservation plan in Alagadi encourages the improvement of the quality of life of residents, the local community expressed negative attitudes due to the perceived economic loss arising from the strategies outlined in the ASPAEP. Similarly, the All-inclusive Optimization Programme's aims at achieving a balanced distribution of economic benefits among stakeholders in the region and to promote local culture remain unachieved due to conflicting interests, limited infrastructure and a wide variety of stakeholders involved in tourism hinder the realization of objectives.

The case of Cyprus confirms that the nature and structure of island tourism challenges the doctrines of sustainable tourism implementation. The intertwined nature of politics on islands, their economic and environmental disadvantage and dependence on external forces for investment and growth pave a difficult road to sustainability. Nevertheless, the challenge of sustainable tourism development in island contexts should neither diminish the merit of sustainability nor divert planners and decision-makers away from the concept's goal. Hitherto, the physical and economic limitations of island destinations have led to the idealized perception of tourism as a vehicle for economic growth. Yet, as literature illustrates, tourism development in islands has a cost. The alignment of tourism development with sustainability principles is imperative for the future of islands. The maturity of several islands in the Mediterranean accelerates the need for reconfiguring developmental approaches, and issues of destination competitiveness, seasonality and quality of life of the local community need be further researched. Thus, in light of this argument, we propose a series of recommendations that could enhance the effort of sustainable island tourism development and direct practitioners' and academics' attention towards the realization that the industry's sustainable development is unlikely to involve a short-term or a complete transformation. Rather, the recommendations presented below should serve as a starting point in what needs to be an on-going, long-term commitment to developing sustainable island tourism.

3.5.1 Recommendations

- Establish clear environmental standards with regard to development and to monitor sustainability growth;
- Provide incentives for the adoption of sustainability practices in tourism by key stakeholders (i.e. tax exemptions, monetary incentives);
- Follow a bottom-up approach in educating tourism stakeholders over the importance of sustainability and environmental conservation;
- Offer monetary incentives to improve entrepreneurship initiative in niche tourism (i.e. rural tourism);
- Provide training to employees in the tourism market to improve quality of service offered and overall competitiveness of destination;
- Educate local community over the importance of sustainability through the inclusion of sustainability principles in schools' and universities' curricula;
- Offer training programmes for rejuvenating traditional crafts and arts;
- Promote inclusive decision-making and community empowerment; and
- Establish a multi-sectoral organization (i.e. members from public, private and non-profit sectors) responsible for managing sustainability projects, coordinating activities and acting as a bridge of communication among stakeholders.

References

Alipour, H. and Kilic, H. (2005) An institutional appraisal of tourism development and planning: the case of the Turkish Republic of North Cyprus (TRNC). *Tourism Management* 26(1), 79–94.

Altinay, M. and Hussain, K. (2005) Sustainable tourism development: a case study of North Cyprus. *International Journal of Contemporary Hospitality Management* 17(3), 272–280.

Andereck, K.L. and Nyaupane, G. (2011) Development of a tourism and quality-of-life instrument. In: Budruk, M. and Phillips, R. (eds) *Quality-of-life Community Indicators For Parks, Recreation and Tourism Management.* Springer, the Netherlands, pp. 95–113.

Andereck, K.L., Valentine, K.M., Vogt, C.A. and Knopf, R.C. (2007) A cross-cultural analysis of tourism and quality of life perceptions. *Journal of Sustainable Tourism* 15(5), 483–502.

Andriotis, K. (2004) The perceived impact of tourism development by Cretan residents. *Tourism and Hospitality Planning & Development* 1(2), 123–144.

Apostolopoulos, Y. and Gayle, D.J. (2002) From MIRAB to TOURAB? Searching for sustainable development in the Maritime Caribbean, Pacific, and Mediterranean. In: Apostolopoulos, Y. and Gayle, D. (eds) *Island Tourism and Sustainable Development: Caribbean, Pacific, and Mediterranean Experiences.* Praeger Publishers, Westport, Connecticut, pp. 3–14.

Bahar, O. and Kozak, M. (2007) Advancing destination competitiveness research: Comparison between tourists and service providers. *Journal of Travel & Tourism Marketing* 22(2), 61–71.

Baldacchino, G. and Ferreira, E.C.D. (2013) Competing notions of diversity in archipelago tourism: transport logistics, official rhetoric and inter-island rivalry in the Azores. *Island Studies Journal* 8(1), 84–104.

Bianchi, R.V. (2004) Tourism restructuring and the politics of sustainability: A critical view from the European periphery (the Canary Islands). *Journal of Sustainable Tourism* 12(6), 495–529.

Briassoulis, H. (2003) Crete: Endowed by nature, privileged by geography, threatened by tourism? *Journal of Sustainable Tourism* 11(2–3), 97–115.

Chapman, A. and Speake, J. (2011) Regeneration in a mass-tourism resort: The changing fortunes of Bugibba, Malta. *Tourism Management* 32(3), 482–491.

Choi, H.C. and Sirakaya, E. (2006) Sustainability indicators for managing community tourism. *Tourism Management* 27(6), 1274–1289.

Crouch, G.I. (2010) Destination competitiveness: An analysis of determinant attributes. *Journal of Travel Research* 49(3), 1–19.

Cuccia, T. and Rizzo, I. (2011) Tourism seasonality in cultural destinations: Empirical evidence from Sicily. *Tourism Management* 32(3), 589–595.

Cucculelli, M. and Goffi, G. (2016) Does sustainability enhance tourism destination competitiveness? Evidence from Italian Destinations of Excellence. *Journal of Cleaner Production* 111, 370–382.

CTO (2001) Annual Report 2001. Available at: http://www.visitcyprus.biz (accessed March 2014).

CTO (2014) Annual Report 2013. Available at: http://www.visitcyprus.biz (accessed May 2015).

Dodds, R. (2007) Sustainable tourism and policy implementation: Lessons from the case of Calviá, Spain. *Current Issues in Tourism* 10(4), 296–322.

Dodds, R. and Butler, R. (2009) Barriers to implementing Sustainable Tourism Policy in Mass Tourism Destinations. *TOURISMOS: An International Multidisciplinary Journal of Tourism* 5(1), 35–53.

Dodds, R. and Kelman, I. (2008) How climate change is considered in sustainable tourism policies: a case of the Mediterranean islands of Malta and Mallorca. *Tourism Review International* 12(1), 57–70.

Dredge, D. (2006) Networks, conflict and collaborative communities. *Journal of Sustainable Tourism* 14(6), 562–581.

Dwyer, L., Forsyth, P. and Spurr, R. (2004) Evaluating tourism's economic effects: new and old approaches. *Tourism Management* 25(3), 307–317.

Farmaki, A. (2015) Regional network governance and sustainable tourism. *Tourism Geographies* 17(3), 1–23.

Farmaki, A., Constanti, P., Yiasemi, I. and Karis, P. (2014) Responsible tourism in Cyprus: the rhetoric and the reality. *Worldwide Hospitality and Tourism Themes* 6(1), 10–26.

Farmaki, A., Altinay, L., Botterill, D. and Hilke, S. (2015) Politics and sustainable tourism: The case of Cyprus. *Tourism Management* 47, 178–190.

Hall, C.M. (2007) *Tourism Planning: Policies, Processes and Relationships*. Prentice Hall, Harlow, UK.

Issa, J.J. and Jayawardena, C. (2003) The 'all-inclusive' concept in the Caribbean. *International Journal of Contemporary Hospitality Management* 15(3), 167–171.

Kerr, S.A. (2005) What is small island sustainable development about? *Ocean & Coastal Management* 48(7), 503–524.

Kim, K., Uysal, M. and Sirgy, M.J. (2013) How does tourism in a community impact the quality of life of community residents? *Tourism Management* 36, 527–540.

Ko, T.G. (2005) Development of a tourism sustainability assessment procedure: a conceptual approach. *Tourism Management* 26(3), 431–445.

Kokkranikal, J., McLellan, R. and Baum, T. (2003) Island tourism and sustainability: A case study of the Lakshadweep Islands. *Journal of Sustainable Tourism* 11(5), 426–447.

Lim, C.C. and Cooper, C. (2009) Beyond sustainability: optimising island tourism development. *International Journal of Tourism Research* 11(1), 89–103.

Liu, Z. (2003) Sustainable tourism development: A critique. *Journal of Sustainable Tourism* 11(6), 459–475.

Loukissas, P. and Triantafyllopoulos, N. (2002) Island-community involvement in tourism land use planning. In: Apostolopoulos Y. and Gayle, D.G. (eds) *Island Tourism and Sustainable Development: Caribbean, Pacific, and Mediterranean Experiences*. Praeger Publishers, Westport, Connecticut, pp. 249–272.

McElroy, J.L. and Albuquerque, K. (2002) Problems for managing sustainable tourism in small islands. In: Apostolopoulos, Y. and Gayle, D.G. (eds) *Island Tourism and Sustainable Development: Caribbean, Pacific, and Mediterranean Experiences*. Praeger Publishers, Westport, Connecticut, pp. 15–31.

Martín, M., María, J., De Dios, J.A. and Molina Moreno, V. (2014) Impacts of seasonality on environmental sustainability in the tourism sector based on destination type: An application to Spain's Andalusia region. *Tourism Economics* 20(1), 123–142.

Mazanec, J.A. and Ring, A. (2011) Tourism destination competitiveness: second thoughts on the World Economic Forum reports. *Tourism Economics* 17(4), 725–751.

Mazanec, J.A., Wöber, K. and Zins, A.H. (2007) Tourism destination competitiveness: from definition to explanation? *Journal of Travel Research* 46(1), 86–95.

Miller, G., Rathouse, K., Scarles, C., Holmes, K. and Tribe, J. (2010) Public understanding of sustainable tourism. *Annals of Tourism Research* 37(3), 627–645.

Nadal, J.R., Font, A.R. and Rossello, A.S. (2004) The economic determinants of seasonal patterns. *Annals of Tourism Research* 31(3), 697–711.

Ritchie, J.B. and Crouch, G.I. (2003) *The Competitive Destination: A sustainable tourism perspective*. CAB International, Wallingford, UK.

Roberts, S. and Tribe, J. (2008) Sustainability indicators for small tourism enterprises – An exploratory perspective. *Journal of Sustainable Tourism* 16(5), 575–594.

Scheyvens, R. (2011) The challenge of sustainable tourism development in the Maldives: Understanding the social and political dimensions of sustainability. *Asia Pacific Viewpoint* 52(2), 148–164.

Swarbrooke, J. (1999) *Sustainable Tourism Management*. CAB International, Wallingford, UK.

Twining-Ward, L. and Butler, R. (2002) Implementing Sustainable Tourism Development on a Small Island: Development and the Use of Sustainable Tourism Development Indicators in Samoa. *Journal of Sustainable Tourism* 10(5), 363–387.

UNWTO (2014) Tourism in small island developing states: Building a more sustainable future for the people of islands. UNWTO, Madrid. Available at: http://www.e-unwto.org/doi/pdf/10.18111/9789284416257 (accessed 4 July 2015).

Yaşarata, M., Altinay, L., Burns, P. and Okumus, F. (2010) Politics and sustainable tourism development – Can they co-exist? Voices from Cyprus. *Tourism Management* 31, 345–356.

Yaw, F. (2005) Cleaner technologies for sustainable tourism: Caribbean case studies. *Journal of Cleaner Production* 13(2), 117–134.

Part II Competitiveness and Quality of Life

4 The Various Faces of Competitiveness in Tourism: A Survey of the Extant Literature

Robertico Croes[1]* and Marketa Kubickova[2]

[1]University of Central Florida, USA; [2]University of South Carolina, USA

4.1 Introduction

This chapter provides a glimpse into the debate concerning tourism competitiveness. While the chapter does not provide a complete review of the literature on this topic, it does present a survey of the issues that inform the current debate on tourism competiveness. The extant tourism competitiveness literature is rather fragmented and dispersed, and spans a wide range of tourist destinations and products. The result of this situation is that current research results do not make for easy comparison, and the existing knowledge on destination competitiveness is still scarce. In light of this, the current chapter sets out to reveal the conceptual and empirical discussion pertaining to tourism competitiveness and the foundational challenges it faces. Furthermore, it will consider a number of research pathways that aim at gaining more meaningful insights into the concept of tourism competitiveness.

The term competitiveness is not a new concept in the tourism literature. Influenced by the works of Porter (1985, 1990), research on tourism competitiveness really took off in the second half of the 1990s with publications by Ritchie and Crouch (1993), Kozak and Rimmington (1999) and Crouch and Ritchie (1999). In addition, two prominent journals dedicated special issues to this subject, namely *Tourism* (1999) and *Tourism Management* (2000). However, the last decade has witnessed heightened attention directed towards this topic in academic circles. Numerous articles have been written on the concept of competitiveness, its measurements and definitions (Buhalis, 2000; Hassan, 2000; Mihalič, 2000; Dwyer and Kim, 2003; Enright and Newton, 2004, 2005; Gooroochurn and Sugiyarto, 2005; Mazanec *et al.*, 2007; Chens *et al.*, 2008; Croes, 2010, 2011; Crouch, 2011; Mazanec and Ring, 2011; Croes and Kubickova, 2013; Assaker *et al.*, 2014; Ayikoru, 2015). International organizations also weighed in, such as the World Economic Forum (WEF) with their travel and

*E-mail: robertico.croes@ucf.edu

tourism competitiveness index (TTCI), the OECD as well as the European Union (EU). Increasingly, national and regional governments have tended to anchor their development strategies and policies on the concept of tourism competitiveness.

Despite this heightened attention, however, tourism competitiveness is sometimes viewed as a 'problematic concept'. While some strides have been made in constructing the theoretical underpinnings of the concept of tourism competitiveness, the ontological and epistemological foundations of the concept remain hazy. The extant tourism literature seems more concerned with searching for factors or determinants of tourism competitiveness and the corresponding statistical techniques than with strengthening the conceptual foundations of tourism competitiveness. For example, Mazanec and Ring (2011) insisted on searching for pathways to improve the WEF TTCI, accepting at face value the definition of tourism competitiveness provided by WEF. Similarly, Benito *et al.* (2014) define the current issues in the tourism competitiveness literature in terms of statistical techniques instead of attempting a more foundational enquiry.

The lack of a more profound conceptual interrogation into the foundations of tourism competitiveness seems to be a consequence of the nature of mainstream literature on competitiveness. For example, mainstream literature reveals diverse perspectives vis-à-vis definition, understanding and measurement (Fagerberg, 1996; Krugman, 1996; Lall, 2001). Defining the concept of competitiveness has therefore become cumbersome. The difficulty defining competitiveness seems to stem from competing perspectives on the usefulness of the very term (Wint, 2003). Management analysts such as Michael Porter are comfortable applying the concept to national economies, while economists such as Paul Krugman question the meaningfulness of the concept to national economies. In the latter's view, competitiveness is a meaningless word when applied to national economies. Furthermore, they feel that the 'obsession' with competitiveness is both wrong and dangerous.

4.2 Defining Competitiveness

In the academic literature, numerous definitions of competitiveness exist today. Examples of these can be found since the Reagan administration in 1984, where competitiveness was defined as 'the degree to which it can, under free and fair market conditions, produce goods and services that meet the test of international markets while simultaneously expanding the real income of its citizens' (Cho and Moon, 1998, p. 12). In a similar vein, Porter (1990, p.1) defines the notion of a competitive strategy as 'a profitable and sustainable position against the forces that determine industry competition'. Aiginger (1996, pp. xiii–xiv) points out that 'a country is said to be competitive if it sells enough products and services, at factor incomes in line with country's (current and constantly changing) aspiration level at macro conditions (of economic and social system) seen as satisfactory by the people'. The WEF defines competitiveness as 'the set of institutions, policies and factors that determine the level of productivity of a country' (World Economic Forum, 2015). These definitions highlight three interrelated components of competitiveness: success (profitability); productivity; and the expansion of income together with the satisfaction of the people.

The tourism literature seems to adhere to these three components. For example, Ritchie and Crouch (2003, p. 2) define competitiveness as the 'ability to increase tourism expenditure, to increasingly attract visitors while providing them with satisfying, memorable experiences, and to do so in a profitable way, while enhancing the well-being of destination residents and preserving the natural capital of the destination for future generations'. Buhalis (2000, p. 106) adds that competitiveness 'is the effort and achievement of long term profitability [for entrepreneurs], above the average of the particular industry within which they operate as well as above alternative investment opportunities in other industries'. The competitiveness of destination, following D'Hauteserre's (2000) definition, can be defined as 'the ability of a destination to maintain its market position and share and/or to improve upon them through time' (p. 23). Dwyer *et al.* (2002, p. 9) assert that tourism competiveness should be coupled with price: 'competitiveness is a general concept that encompasses price differentials coupled with exchange rate movements, productivity levels of various components of the tourist industry and qualitative factors affecting the attractiveness or otherwise of a destination'.

Heath (2003, p. 9) posited that destination competitiveness 'includes objectively measured variables such as visitor numbers, market share, tourist expenditure, employment, value added by the tourism industry, as well as subjectively measured variables, such as 'richness of culture and heritage,' 'quality of the tourism experience, etc.' Enright and Newton (2004, p. 778) propound that 'a destination is competitive if it can attract and satisfy potential tourists and this competitiveness is determined both by tourism-specific factors and by a much wider range of factors that influence the tourism service providers'. We find mention of similar elements within the definition of Hong (2009): 'the competitive position (with high profits and constant growth) of the tourism industry of a nation relative to the global market of tourist industries in other nations, whether developed or developing countries, which therefore increases the real income and standard of living of its citizens.' Azzopardi (2011, p. 22) defines tourism competitiveness in the context of destinations as 'the ability of the destination to identify and exploit comparative advantages and enhance competitive advantages to attract visitors to a destination by offering them a unique overall experience for a fair price that satisfies the profit requirement of the industry and its constituent elements, as well as the economic prosperity objective of the residents, without jeopardizing the inalienable aspirations of future generations'.

Other studies have defined destination competitiveness either explicitly or implicitly as having 'more of' something – for example market share, profits, success and so on, than another destination has. For example, the studies of Hassan (2000), Sahli (2006) and Craigwell (2007) suggest that destinations that enjoy a greater number of arrivals and more spending from tourists, or ones that have benefited from a higher market share in the global market than others have, are considered to be competitive destinations. There is, clearly, a widely held view that competitiveness should be linked to high visitor numbers and increasing destination income.

However, these definitions are anchored on two seemingly contradictory frameworks of analysis, namely the comparative advantage school and the strategic management school. For example, a review of the literature from Adam Smith to Michael Porter reveals that the term 'competitiveness' has evolved from the conditions of resource availability and technology (i.e. a comparative advantage) to the deployment of resources adjusting to customer preferences (i.e. a competitive advantage). Assessing the endowment of natural resources and the productivity differences between countries is the basis for comparison from the comparative advantage perspective, and it also explains trade patterns. In other words, costs differentials are considered as the basis for international trade. For example, the availability of accessible natural resources (such as beaches, sunny weather, or natural areas) becomes a comparative advantage in the tourism product function and explains why destinations with these elements have specialized in tourism. On the other hand, management theories, as the one espoused by Porter, posit that an advantage exists in creating core competencies deploying and using resources effectively; this is achieved by applying resources to customer preferences (i.e. the deployment of resources). The application of resources is therefore more important than inherited resources when one sets out to explain wealth and trade (Smit, 2010).

The discussion above reveals four components that together constitute tourism competiveness: (i) the ability to deploy resources; (ii) the ability to spawn memorable experience of tourists; (iii) the ability to have superior performance; and (iv) the ability to enhance quality of life. In other words, tourism competitiveness seems to be concerned with factors (such as infrastructure, people and technology) leading to long-term performance (potential); also with the process that transforms this potential into achievement; and with making the achievement sustainable (performance). Table 4.1 provides a summary of definitions pertaining to competitiveness.

Table 4.1. Selected definitions of tourism competitiveness by year of publication.

Source (year)	**Definition**
Reagan administration (1984) in Cho and Moon (1998, p. 12)	'the degree to which it can, under free and fair market condition, produce goods and services that meet the test of international markets while simultaneously expanding the real income of its citizens'
Porter (1990, p. 1)	'a profitable and sustainable position against the forces that determine industry competition'
OECD (1994, p. 18)	'the degree to which a nation can produce goods and services which meet the test of international markets, while simultaneously maintaining or expanding the real income of its people over the long term.'
Aiginger (1996, pp. xiii–xiv)	'a country is said to be competitive if it sells enough products and services, at factor incomes in line with country's (current and constantly changing) aspiration level at macro conditions (of economic and social system) seen as satisfactory by the people'
Crouch and Ritchie (1999, p. 137)	'the ability of destinations to provide a high standard of living for residents of the destination'
Hassan (2000, p. 240)	'destinations' ability to create and integrate value-added products that sustain its resources while maintaining market position relative to competitors'

continued

Table 4.1. *Continued*

Source (year)	Definition
Buhalis (2000, p. 106)	'effort and achievement of long term profitability, above the average of the particular industry within which they operate as well as above alternative investment opportunities in other industries'
D'Hauteserre (2000, p.23)	'the ability of a destination to maintain its market position and share and/or to improve upon them through time. To some extent competitiveness also means an extended product life time'
Crouch and Ritchie (2000, p. 2)	'what makes a tourism destination truly competitive is its ability to increase tourism expenditure, to increasingly attract visitors while providing them with satisfying, memorable experiences, and to do so in a profitable way, while enhancing the well-being of destination residents and preserving the natural capital of the destination for future generation'.
Dwyer *et al.* (2002, p. 9)	'competitiveness is a general concept that encompasses price differentials coupled with exchange rate movements, productivity levels of various components of the tourist industry and qualitative factors affecting the attractiveness or otherwise of a destination.'
Ritchie and Crouch (2003, p. 2)	'[the] ability to increase tourism expenditure, to increasingly attract visitors while providing them with satisfying, memorable experiences, and to do so in a profitable way, while enhancing the well-being of destination residents and preserving the natural capital of the destination for future generations'
Heath (2003, p. 9)	'includes objectively measured variables such as visitor numbers, market share, tourist expenditure, employment, value added by the tourism industry, as well as subjectively measured variables, such as "richness of culture and heritage," "quality of the tourism experience," etc.'
Enright and Newton (2004, p. 778)	'a destination is competitive if it can attract and satisfy potential tourists and this competitiveness is determined both by tourism-specific factors and by a much wider range of factors that influence the tourism service providers'
Bristow (2005, p. 129)	'The relative competitive position (in terms of profits and growth) of a nation's tourism industry in the global market, including developed and developing countries, which could therefore increase the real income of its citizens and improve its standard of living.'
Hong (2009, p. 37)	'the competitive position (with high profits and constant growth) of the tourism industry of a nation relative to the global market of tourist industries in other nations, whether developed or developing countries, which therefore increases the real income and standard of living of its citizens.'

continued

Table 4.1. *Continued*

Source (year)	Definition
Kayar and Kozak (2010, p. 4)	'the set of institutions, policies, and factors that determine the level of productivity of a country involving static and dynamic components'
Azzopardi (2011, p. 22)	'Tourism competitiveness can be defined as the ability of the destination to identify and exploit comparative advantages and enhance competitive advantages to attract visitors to a destination by offering them a unique overall experience for a fair price that satisfies the profit requirement of the industry and its constituent elements, as well as the economic prosperity objective of the residents, without jeopardizing the inalienable aspirations of future generations.'
World Economic Forum (2015)	'as the set of institutions, policies and factors that determine [sic] the level of productivity of a country. The productivity, in turn, sets the level of prosperity that can be earned by an economy'

4.3 The Construct of Competitiveness

According to the study of Croes and Kubickova (2013), the construct of competitiveness is predicated on three important assumptions: (i) it is embedded in the concept of competition; (ii) it is a multi-dimensional construct; and (iii) it occurs on different levels. These assumptions are very important when determining competitiveness in tourism. In light of these, competitiveness is not only challenging to define, but also difficult to measure. Each of these assumptions are further discussed in the next sections (Vlami *et al.*, 2006).

First, the construct of competitiveness is embedded in the concept of competition: 'to compete is to consciously attempt to gain an advantage, or to defend and maintain a position, in relation to adversaries' (Leiper, 2008, p. 242). In other words, competition can be translated as 'rivalry' (Porter, 1990) or a zero sum game (if one gains, another loses). The idea of a zero sum game, according to the viewpoint of the 16th-century mercantilists, can be achieved by means of higher exports and restricted imports, resulting in an inflow of gold and silver, and thus making a country rich and powerful (Smit, 2010). Competition, however, can also refer to the role of information which determines the working of the market (perfect competition or market failure). There are three factors that seem to prompt increased competition. The first of these is lodged in the number of destinations (Croes and Kubickova, 2013). Over the years, the number of destinations has increased while at the same time the number of original markets remained the same, thus increasing competition (Vlami *et al.*, 2006; Croes and Rivera, 2010). A second reason for increased competition is that destinations have become easily substitutable. In other words, if other destinations provide similar experiences to ones that are readily accessible, the alternatives are likely to be chosen (Mangion *et al.*, 2005; Pike, 2005; Lew and McKercher, 2006). As Pike (2005) points out, today's travellers are spoiled for choice of available destinations, and this has a bearing on their decision-making (Lew and McKercher, 2006). A third reason for the growth of competitiveness is that growth rates of tourism demand have fallen significantly over the past 50 years, despite increased demand in tourism globally (Papatheodorou and Song, 2005). Therefore, determining and understanding the level of international competitiveness is an ever-growing concern for governments and firms (Smit, 2010). In particular, those countries that are relying on tourism for national income are more than ever interested in gaining competitive advantage (Gooroochurn and Sugiyarto, 2005).

In the second place, the construct is multi-dimensional and involves a number of attributes (Crouch and Ritchie, 1999; Dwyer and Kim, 2003; Enright and Newton, 2004, 2005; Gooroochurn and Sugiyarto, 2005; Craigwell, 2007; Wang *et al.*, 2012). Some authors have set out to assess competitiveness by utilizing the level of corruption in the destination or by number of arrivals, room nights and value added. However, the literature is not clear in terms of how these attributes are related. For example, can we say that if a destination has beautiful natural resources, it is more competitive than one that lacks such resources? Just because a destination has more hotel rooms, this does not automatically translate into more arrivals and thus into greater competitiveness. It follows that a destination's attributes are viewed as inputs that are somehow mysteriously mixed in a 'black box' that leverages a desired societal outcome. This view of an input framework is predicated on the potential of a destination to accomplish its developmental objectives without accounting for intervening factors such as inequality and institutional weaknesses that can detract from the quality of life of residents of a destination. Intervening factors, however, are seen as pervasive in the context of developing countries. In the previous section, we identified the interconnectedness of three dimensions, i.e. potential, process and performance, embedded in the tourism competitiveness construct.

In the third place, one can argue that competitiveness occurs on different levels: on firm, region and national levels (Wang *et al.*, 2012). Therefore, due to the different assumptions and levels (firm, industry, nation), various definitions of competitiveness exist today, but none of these are widely accepted (Krugman, 1994; Ritchie and Crouch, 2003; Berger, 2008). For example, Michael Porter feels comfortable applying the notion of competitiveness to nations, while Paul Krugman questions the usefulness of the term in the context of nations (Croes, 2011). Several studies have assumed that the nature of competitiveness at the firm and national level is identical. However, unlike at the firm level, a falling market share in tourism does not mean a loss of national competitiveness. What it suggests, rather, is a shift in comparative advantage (Krugman, 1996; Lall, 2001; Croes and Rivera, 2010). Thus, what can be applied to firms cannot be directly applied to destinations and vice versa (Ritchie and Crouch, 1993; Croes and Rivera, 2010; Croes, 2011; Wang *et al.*, 2012).

The notion of tourism competitiveness also reveals four characteristics that can be gleaned from the tourism literature. Tourism competitiveness implies long-term performance that is assessed based on productivity levels and value-added growth at a destination over time. Furthermore, the term also refers to the measure of control that a destination has over its resources and assets (controllability) such as quality of the product, marketing and information technology. How a destination is able to control its resources in order to attract tourists as compared to other destination reveals the third characteristic of tourism competitiveness, namely relativity. Finally, tourism competitiveness is dynamic in nature, constantly evolving through building capabilities and producing new offerings. This dynamic nature is best exemplified in the tourism area life cycle (TALC) framework.

4.4 Measurements of Competitiveness

Measuring tourism competitiveness has become as problematic as trying to define it. Conventionally, tourism competitiveness has been measured by means of various indices that have been developed by different researchers and organizations, and utilizing different indicators. As Gooroochurn and Sugiyarto (2005, p. 26) purport: 'identifying the elements of competitiveness is contentious because of the conceptual problem embodied in its definition. Competitiveness is a relative concept, and its measure will vary depending on the choice of the base year and/or the base country'.

Because competitiveness cannot be measured directly, studies have used various methods, models and indicators to measure competitiveness, depending on both tourism-specific factors and general business factors, and involving both the micro- and the macro-environments (Das and DiRienzo, 2010; Kayar and Kozak, 2010). For example, Dwyer and Kim (2003) associated competitiveness with three major groups of thought. The first group of thought adheres to the comparative advantage and/or price competitiveness perspective, where economists emphasized price and country-specific characteristics. The second group of thought belongs to a strategy and management perspective, and they focus on firm-specific characteristics (micro-perspective). Finally, the third group of thought holds a historical and socio-cultural perspective that centres on a range of social, cultural and political characteristics (macro-perspective) (Dwyer and Kim, 2003).

Using a different lens, Kozak and Rimmington (1999) point out that, in general, competitiveness can be defined from the point of view of input and output. The input side is based on physical and human capital endowment (such as land, parts of the infrastructure, transport, hotel services, etc.), while the output side involves profitability, market share, growth, productivity and so on. Others clearly imply competition with other destinations, basing it on arrival flows, bed-nights, value-added and customer satisfaction (Croes and Kubickova, 2013). A comprehensive summary of variables used in the tourism literature is presented in Table 4.2.

These different measures used to determine tourism competitiveness are not without a measure of vagueness. For example, a study by Dwyer *et al.* (2000b) suggests that *price* is the factor affecting competitiveness. Based on this logic, it can be argued that an increase in labour cost or a sudden appreciation of a destination's currency would lead to a decline in competitiveness. However, in reality – as Croes (2011) argues – empirical evidence implies otherwise; this is because market share for export (tourism) and relative unit costs (price) tend to move together in the so-called Kaldor paradox. Apart from price, other researchers and institutions often utilize *inputs* to measure competitiveness. Again, based on this logic it can be argued that to have resources (inputs) would automatically translate into a success (output). In other words, just because two destinations have the same resources, does it mean they are both equally competitive? Obviously, this is not the case.

Table 4.2. Variables/constructs used in measuring tourism competitiveness.

Author(s) (year)	**Variable(s) used**
De Keyser and Vanhove (1994)	Macroeconomic factors, factors on the supply and demand side, transport factors, tourism policy-related factors
Crouch and Ritchie (1999)	Economic, social, environmental conditions
Kozak and Rimmington (1999)	Qualitative (e.g. likes and dislikes regarding the destination) and quantitative (tourist numbers, tourism revenues)
Dwyer *et al.* (2000b)	Travel cost to and from 19 destinations
Mihalič (2000)	Environmental quality
Crouch and Ritchie (2000)	Core resources and attractors, supporting factors and resources, destination management, qualifying determinants, amplifying and qualifying determinants, destination policy, planning, development
Go and Govers (2000)	Facilities, accessibility, service quality, image, climate, environment, attractions
Hassan (2000)	Comparative advantages, demand-orientation, tourism sector structure, environmental care
Dwyer *et al.* (2000b)	Price
Enright and Newton (2004)	Business-related factors, image, attractiveness
Dwyer *et al.* (2002)	Exchange rate
Dwyer and Kim (2003)	Inherited resources, created resources, supporting resources, demand conditions, situational conditions, destination management
Dwyer *et al.* (2004)	Destination management, nature-based resources, heritage resources, quality service, efficient public services, tourism shopping, government commitment, location and access, e-business, night life, visa requirements, amusement/theme parks
Enright and Newton (2004, 2005)	Tourism attractors (safety, cuisine, visual appeal, nightlife, climate, museum and galleries, etc.), business-related factors (political stability, government policy, staff cost, level of technology, access to information, etc.)
Guo and Cao (2004)	Core attractors, basic resources, supporting factors, developing factors, qualifying factors, management innovation
Gooroochurn and Sugiyarto (2005)	Price, economic and social impact, human resources, infrastructure, environment, technology, openness, social development
Mazanec *et al.* (2007)	Price competitiveness (based on exchange rate adjusted purchasing power parities)
Zang and Huang (2007)	Destination management, core resources and attractions, qualifying and amplifying factors, supporting factors
Li *et al.* (2007)	Tourism resources, tourism facilities, tourist service, internal environment

continued

Table 4.2. *Continued*

Author(s) (year)	**Variable(s) used**
Song (2008)	Attraction, environment, tourism product, tourism marketing
Gursoy *et al.* (2009)	Price competitiveness, environment, technology, openness, social development, human development.
Mathew (2009)	Support activities, key success drivers, inputs, sustainable process, leverage/operation, destination marketing, sustainable services
Kayar and Kozak (2010)	Policy rule and regulation, environmental regulation, safety and security, health and hygiene, prioritization of travel and tourism, air and ground transport infrastructure, tourism infrastructure, information and communication technology infrastructure, price competitiveness, human resources, national tourism perceptions, natural and cultural resources
Jackman *et al.* (2011)	Comparative advantage (Balassa index)
Chen *et al.* (2011)	Decision-making process (purposes of trip, duration of trip, information resources, time of decision-making, decision-makers), tourists' destination image before and after the trip, overall satisfaction with destination and service performance and the willingness to revisit and recommend
Kim *et al.* (2011)	Accessibility, transportation, attractions, cost, facilities and services, availability, staff services, convention and exhibition facilities, climate, safety
Gomezelj Omerzel (2011)	Inherited resources, created resources, supporting factors, situational conditions, management, demand
Wang *et al.* (2012)	Destination management, tourism resources, tourism superstructure and infrastructure, destination-supporting factor

To measure competitiveness, furthermore, researchers have utilized different research methods. Some have employed quantitative measures (Kozak and Rimmington, 1999; Dwyer *et al.*, 2004; Enright and Newton, 2004, 2005; Gooroochurn and Sugiyarto, 2005; Gursoy *et al.*, 2009; Das and DiRienzo, 2010) utilizing primary and/or secondary data. Others have set out to determine competitiveness by utilizing qualitative measures (D'Hauteserre, 2000; Hassan, 2000; Kaynak and Marandu, 2006) and some have combined qualitative and quantitative measures (Kozak and Rimmington, 1999; Hudson *et al.*, 2004; Campos-Soria *et al.*, 2005).

4.4.1 Three areas of research

A review of the tourism literature reveals three obvious spheres of research (Crouch, 2011). The first group of researchers has concentrated on diagnosing positions of specific destinations (case studies), such as Spain and Turkey (Kozak and Rimmington, 1999; Kozak, 2003), Turkey (Icoz *et al.*, 1998), European cities (Mazanec *et al.*, 2007), the Caribbean (Croes, 2011), Central America (Croes and Kubickova, 2013), or Las Vegas (Chon and Mayer, 1995). Additional competitiveness studies that concentrate on destinations are listed in Table 4.3.

Table 4.3. Competitiveness studies analysing specific destination(s).

Author (year)	Destination
Ahmed and Krohn (1990)	USA
De Keyser and Vanhove (1994)	The Caribbean
Chon and Mayer (1995)	Las Vegas
Mazanec (1995)	European cities
Pearce (1997)	South-east Asia
Faulkner *et al.* (1999)	South Australia
Kim *et al.* (2000)	South Africa
Carmichael (2002)	Toronto, Canada
Papatheodorou (2002)	Mediterranean resorts
Dwyer *et al.* (2003)	Australia
Enright and Newton (2004, 2005)	Asia Pacific
Hudson *et al.* (2004)	Canadian ski resorts
Chens *et al.* (2008)	Cambodia
Miller *et al.* (2008)	Cuba
Gursoy *et al.* (2009)	Middle East
Croes and Rivera (2010)	Puerto Rico
Kayar and Kozak (2010)	EU and Turkey
Chen *et al.* (2011)	Kinmen National Park in Taiwan
Kim *et al.* (2011)	East Asia
Jackman *et al.* (2011)	Small island developing states
Schalber and Peters (2012)	The Alps
Croes and Kubickova (2013)	Central America
Mingaleva and Bunakov (2014)	Russia

The second group of studies has attempted to assess the determining factor/constructs of competitiveness; for example, access to tourist markets, prices, cost or currency exchange rates (Icoz *et al.*, 1998; Kozak and Rimmington, 1999), the environment (Hassan, 2000; Mihalič, 2000), strategic management (Jamal and Getz, 1996; Soteriou and Roberts, 1998; Evans *et al.*, 2002), quality management (Go and Govers, 2000), including destination management systems (Baker *et al.*, 1996) and destination marketing (Buhalis, 2000). Additional factors utilized in competitive research are listed in Table 4.4.

Finally, the third group of studies set out to develop models and thus to provide a theoretical foundation for competitiveness that is not specific to particular destination or attributes (Crouch, 2011). In the academic literature, seven well-established indices have been identified and extensively discussed by a number of researchers. Apart from these well-known published rankings (shown below), there are many unpublished ones that have been prepared by consultants, governments and research institutions (Lall, 2001).

Table 4.4. Factor(s) utilized in tourism competitive research.

Factor(s)/constructs(s)	Author(s)
Planning methods and price competitiveness	Stevens (1992), Dwyer *et al.* (2000a, b, 2001, 2002), Saayman and Saayman (2008), Dwyer and Forsyth (2011)
Corruption	Das and DiRienzo (2010)
Nature-based tourism	Huybers and Bennett (2003)
Destination positioning	Chacko (1996)
Sustainable tourism/environment	Hassan (2000), Mihalič (2000), Lee and Chen (2010)
The impact on local development	Cunha and Cunha (2005)
Attractiveness/image	Chon *et al.* (1991), Hu and Ritchie (1993), Enright and Newton (2004, 2005)
Transportation	De Keyser and Vanhove (1994), Kim *et al.* (2011)
Supporting factors	Gomezelj Omerzel (2011), Wang *et al.* (2012)

Conceptual Tourism Competitiveness Model

The first index discussed is the Conceptual Tourism Competitiveness Model as developed by Crouch and Ritchie (1999). Their elaborate model is based on Porter's (1990) diamond of the national competitiveness framework discussed in an earlier section. This model includes five distinctive resources that give rise to destination competitiveness and sustainability. These resources are: (i) supporting factors and resources (e.g. destination infrastructure, accessibility, political will, hospitality and resources facilitation); these are a building pillar to (ii) core resources and attractors (e.g. physiography and climate, culture and history, entertainment, and mix of activities); (iii) destination management (e.g. marketing, service, visitor management, etc.); (iv) destination policy, planning and development (e.g. vision, development, audit, philosophy/values, etc.); and (v) qualifying and amplifying determinants (e.g. location, interdependencies, awareness, carrying capacity, etc.); these are influenced not only by the micro-environment of tourism competitiveness (such as human and knowledge resources, capital resources), but also by the global (macro) environment (growth and development, effectiveness). According to the model, competitiveness can be measured either in subjective or objective terms. Subjective measures would be those that relate to visitor perceptions of the natural resources in a destination. An example of objective measures, on the other hand, would be the use of natural resources to indicate the size of areas devoted to nature reserves, mean temperatures and the number of coral reefs (Kaynak and Marandu, 2006).

Destination Competitiveness

The second index that has been proposed is the Destination Competitiveness model developed by Dwyer and Kim (2003). This model is modified from a similar one by Crouch and Ritchie (1999). It explicitly recognizes demand conditions as an important determinant of destination competitiveness. The model also points out that competitiveness is not the ultimate goal; rather, it is an intermediate goal toward the prosperity of a destination and residents' well-being. The model comprises resources (inherited and created), situational conditions (e.g. economic, social, cultural, demographic), destination management factors (e.g. destination marketing, management, destination policy, planning and development, human resources and environmental management), demand conditions (e.g. preferences) and socio-economic prosperity.

Tourism Competitiveness Monitor

The third index is the Tourism Competitiveness Monitor that was developed by Gooroochurn and Sugiyarto (2005). Their index is also based on the Crouch and Ritchie (1999) model that utilizes a wide range of relevant indicators. The main difference between these two studies is the substitution of historical and cultural resources (which are difficult to measure) with a social development indicator, in other words, capturing the quality of life in the society. The Tourism Competitiveness Monitor includes eight tourism indicators: price competitiveness, infrastructure development, environment, technology advancement, human resources, openness, social development and human tourism indicator. The major contribution of the index is that different weights are attached to each indicator, thus demonstrating their relative importance in the aggregate index.

Global Competitiveness Index

The fourth index dealt with here is the Global Competitiveness Index (GCI) developed by World Economic Forum (WEF). Over the past three decades, the goal of this report has been to stimulate discussion and to provide insight in terms of destination competitiveness. The index is also based on Porter's study on competitiveness and it aims to measure those factors that contribute towards the future growth of an economy. The index involves static and dynamic components, and is composed of the 12 pillars of competitiveness. These 12 pillars are then further divided into three sub-pillars: (i) basic requirements (institutions, infrastructure, macroeconomic environment, health and primary education); (ii) efficiency enhancers (higher education and training, goods market efficiency, labour market efficiency, financial market development, technological readiness, market size); and (iii) innovation and sophistication (business sophistication, innovation). As is the case with Porter's model, the pillars are not independent and can reinforce each other. Furthermore, as with Gooroochurn and Sugiyarto, each sub-index is attributed different weight based on GDP per capita (World Economic Forum, 2015).

World Competitiveness Scoreboard

The fifth model discussed here is the World Competitiveness Scoreboard (Yearbook) (WCY) developed by the International Institute for Management Development (IMD). The IMD creates a competitive report based on combination of soft and hard data, grouped into four main factors, which are then further divided into 20 sub-factors. The four main factors are: (i) economic performance; (ii) government efficiency; (iii) business efficiency; and (iv) infrastructure. Each sub-factor has the same weight in the overall consolidation of results, comprising more than 300 criteria altogether into a single index (IMD, 2013).

Travel and Tourism Competitiveness Report

The sixth index dealt with here is the Travel and Tourism Competitiveness Report (TTCI) developed by the World Economic Forum (WEF, 2013). The TTCI is the most popular tool to rank countries in term of their performance, covering over 140 economies in 2013. The goal of this index is to provide a comprehensive strategic tool with the intention of improving and developing the tourism sector in different countries. It is also based on Porter's work, and its aim is to measure microeconomic competitiveness across countries (Lall, 2001). The TTCI is based on three broad categories of variables that are summarized into three sub-indices: (i) the travel and tourism (T&T) regulatory framework sub-index; (ii) the T&T business environment and infrastructure sub-index; and (iii) the T&T human, cultural and natural resources sub-index. Each of these sub-indices is composed of an additional 14 pillars, which are, in turn, composed of a number of individual variables. The dataset includes both qualitative and quantitative data from publicly available sources.

The Croes Tourism Competitiveness Index

Finally, the Tourism Competitiveness Index that has been developed by Croes (2010) allows for timely, accurate and accessible data ranking. The difference between the previous indices listed and the TCI suggested by Croes (2010) is the shift from inputs to outputs; this allows for comparisons of tourism performance to be conducted over time. In addition, the model is characterized by simplicity, by the wide-ranging effects of the indicators it takes into account and by the minimum data requirements it presents. Rather than using the propensity of variables without theoretical justification of their causal relationship to the dependent variable, the TCI comprises three outputs, each portraying different aspects of the industry's productivity. The first output reflects the *current* performance in the global tourism market scale by size; the second output represents *dynamism* of performance over time (growth rate) and the third output characterizes the size of the industrial base in the economic structure. Tourism receipts per capita, average tourism receipts growth rates, and tourism added values as percentage of the GDP were used as variables representing these three outputs (Croes, 2011). The three variables are weighted based on their correlations rather than on equal parts, given the interest in current performance and the need to account for the size of each country. The three outputs, according to the Croes model, also make use of different aspects of the industry's productivity by means of the concept of utility, in this manner allowing for the creation of a quick snapshot of the destination's position in terms of competitiveness.

4.5 Critique of Competitive Indices

All of the tourism competitive models currently available in the academic literature and discussed in the above section have been subjected to some type of flaw in one way or other. Due to the characteristics and complexity of the tourism industry, it is extremely difficult to create a competitive index that addresses all relevant issues. Models tend to be criticized for their complexity, the amount of data needed for analysis, or because of measurement issues, or because there is a lack of clear links between variables. Other criticisms include the models' applicability to destinations, particularly to developing countries. As Dwyer and Kim (2003, p. 407) state:

> the review of the literature on tourism destination competitiveness revealed that none of the models that have been proposed to date are entirely satisfactory . . . they do not provide a comprehensive treatment of the various issues surrounding the notion of 'competitiveness' that are being explored in the wide literature and that must be taken into account in developing a comprehensive framework of destination competitiveness.

One frequent criticism noted in the academic literature is a model's usability (or lack thereof); here one can refer to the Crouch and Ritchie (1999) competitive model. The complexity of the model prevents it from being empirically tested. As Kaynak and Marandu (2006, p. 231) point out, it is extremely difficult to 'compare competitiveness of destinations in respect of all the determinants taken together, as well as their ability to gauge competitiveness in respect of the constituent dimensions of the model'. Another limitation inherent in Crouch and Ritchie's model is its contextual relevance. For example, Ayikoru (2015) contests the universal applicability of the Crouch and Ritchie model – referring to the case of Uganda; the author asserts that 'the model in its "universal" sense does not adequately account for how situational conditions (or chance events) might, for instance, influence a nontraditional destination's quest for competitiveness.' Previously, Beeton (2005, p. 295) expressed another concern regarding this model, 'Ritchie and Crouch focus on a tourism specific model that appears to come primarily from research undertaken in developed countries', adding that 'it would be interesting to test [it] in developing countries' (p. 296).

Another issue often facing these indices is the relative importance of all the factors included in the model. In general, not all factors should have an equal measure of influence on a destination's competitiveness. Some may, obviously, be more important than others. Therefore, attaching the same relative importance to all factors in the overall model is a flaw that is often overlooked (Kaynak and Marandu, 2006). For example, in the WCY index, equal weight is assigned to all variables. However, one can hardly suggest that having natural resources is as important as providing safety and security.

Many indices, such as the TTCI, WCY, or Crouch and Ritchie's model, require a collection of a myriad of indicators, ranging from public finance to business management practices and attitudes and values. In particular, the WCY index requires over 300 factors and criteria. In this sense, the index is not very practical as appropriate proxies for some of the variables are not always available for all destinations (Craigwell, 2007). Therefore, some countries might simply be unable to have all obligatory statistics to provide proper ranking, thus causing themselves to be omitted from the rankings all together. Table 4.5 provides a summary of selected competitiveness indicators.

Certain indicators, such as the TTCI, have been criticized for not demonstrating a sufficiently clear understanding of cause and effect relationships between variables and competitiveness (Croes, 2010). Since various indices are based on literally hundreds of indicators, there is no direct connection between these indicators and the competitive level (outcome) of the countries (Mazanec *et al.*, 2007). The base assumption of these indices is the oversimplified notion that inputs reflect outputs.

Table 4.5. Selected studies and tourism competitiveness indicators (Croes, 2011).

Author/institution	Amount of indicators	Weighted
WEF Growth Competitiveness Index	174	Yes
IMD World Competitiveness Index	143	No
Environmental Sustainability Index	68	No
The Travel and Tourism Competitiveness Index	58	No
Crouch and Ritchie	250	No
Dwyer *et al.*	83	No
Gooroochurn and Sugiyarto	54	Yes
Hong	68	Yes

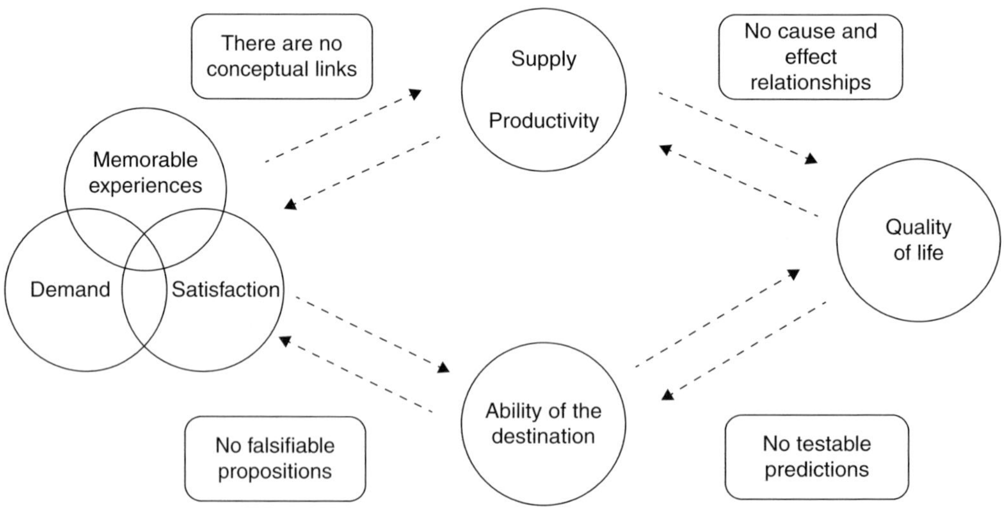

Fig. 4.1. Shortcomings in the Tourism Competitiveness Framework (Croes, 2011).

For example, if a destination has a good infrastructure, this determinant is supposed to predict its attractiveness, resulting in increased arrivals, and thus the destination is rated as more competitive than others. Such nexus of inputs–outputs is not an automatic truth for a number of countries, due to, for example, market imperfections that are amplified by institutional weaknesses (Lall, 2001; Croes, 2010). In addition, such variables fail to take into consideration the nature of the destination (the size of the market), the stage of economic development (the destination life cycle) and the degree of dependency on tourism (Croes, 2011). Finally, the tourism literature lacks clear propositions (Mazanec *et al.*, 2007; Croes, 2011) as well as sufficient empirical testing. Figure 4.1 summarizes the theoretical challenges in the extant tourism competitiveness models.

4.6 Aligning Productivity, Satisfaction and Quality of Life in an Integrated Framework

Tourism competitiveness theory has yet to overcome challenges related to its conceptualization and measurement. Three factors seem to circumscribe this challenge: (i) the Ricardian theory of comparative advantage and the Porterian framework of competitive advantage; (ii) the debate pertaining to measurements skewing towards the application of indicators that encompass inputs (such as land, parts of the infrastructure, transport and hotel services, etc.); and (iii) studies on tourism that imply an automatic direct nexus inputs–output without testing the validity of this nexus.

The proposed new Tourism Competitiveness Integrated Framework (TCIF) considers the multidimensional approach of the construct. It takes into consideration the three dimensions of potential, process and performance together with the four characteristics of long-term view, controllability, relativity and dynamism. Based on these guidelines, the tourism competitiveness construct is predicated on productivity, human agency, satisfaction and quality of life (Croes and Kubickova, 2013). Productivity is associated with factors of production and their efficient configuration, thereby generating value. Productivity is a source of attraction and satisfaction over time, and is captured through tourism growth and value added to the tourism product. When attractiveness and satisfaction combine to accrue visitors' expenditures, the destination manifests its ability to compete. Expenditures are therefore what matters, according to Ritchie and Crouch (2003), implying that arrivals are secondary to expenditures. This perspective considers a destination to be competitive only when such a destination can convert a number of relevant factors into tourism revenues (Li *et al.*, 2013).

Linking productivity to quality of life, in another scenario, cannot explain by itself why some destinations prosper while others fall behind. For example, tourism receipts cannot be considered in isolation to quality of life. Destination managers could consider maximizing revenues in the context of, for example, covering fiscal shortfalls without any consideration of the effects of revenues on quality of life. Maximizing revenues could lead to perverse actions, however. One could think of a destination that is able to maximize revenues while not being able to transpose those revenues to enhance the quality of life of its citizens. For example, Croes (2012) found that while Costa Rica was able to increase its tourism revenues, the destination could not establish a positive relationship between revenues and enhanced quality of life. Blake *et al.* (2006) assessed tourism productivity in the UK; they concluded that the gains from productivity do not wholly accrue to domestic residents.

The missing link in these models is the importance of human agency in deciding how to access and use resources. The access and use of resources constitutes an important condition in the social context – it entails the freedom to achieve as well as the achievement (Sen, 1999). Human agency refers to the ability of an individual to change his or her behaviour and to create a demand for new behaviour to improve and to sustain competitiveness. Human agency thus intervenes in the process of converting resources into quality of life by focusing on the conversion process itself. The conversion of these factors in enhancing or hampering quality of life then relies on human agency.

It is our contention that quality of life should, clearly, be an integral part of any model measuring tourism competitiveness. Therefore, tourism competitiveness exists only under the condition of there being a positive relationship between productivity, satisfaction and quality of life. In other words, a destination is competitive when increasing revenues induce enhanced quality of life. It follows that understanding the conditions under which some destinations perform better than others is crucial for policy makers if tourism is to fulfil its promise of enhancement of quality of life. Future research should identify the conditions under which productivity gains will or will not benefit domestic residents; which barriers are preventing productivity gains from reaching domestic residents; and what the channels are through which these productivity gains will reach domestic residents.

4.7 Conclusions

This survey of the tourism competitiveness literature provided insights on the definition, meaning and measurement of the tourism competitiveness construct. The survey reveals that the foundational underpinnings of the construct are not wholly defined – and this lack of proper definition hampers the theoretical advancement of the tourism competitiveness framework. Sifting through the multiple definitions, the study identified three dimensions and four characteristics revealed in four interrelated components of tourism competiveness; these are productivity, human agency, satisfaction and quality of life. These four components are anchored in the ontological premises of trade theory, consumer choice and human welfare.

References

Ahmed, Z.U. and Krohn, F.B. (1990) Reversing the United States' declining competitiveness in the marketing of international tourism: A perspective on future policy. *Journal of Travel Research* 29, 23–29.

Aiginger, K. (1996) Creating a dynamically competitive economy. In: Devine, P. and Katsoulacos, Y. (eds) *Competitiveness, Subsidiarity And Industrial Policy.* Routledge, Hoboken, New Jersey, pp. 121–146.

Assaker, G., Hallak, R., Vinzi, V.E. and O'Connor, P. (2014) An empirical operationalization of countries' destination competitiveness using partial least squares modeling. *Journal of Travel Research* 53, 26–43.

Ayikoru, M. (2015) Destination competitiveness challenges: A Ugandan perspective. *Tourism Management* 50, 142–158.

Azzopardi, E. (2011) The international competitiveness of Malta as a tourist destination. Doctoral thesis, Robert Gordon University, Aberdeen.

Baker, M., Hayzelden, C. and Sussmann, S. (1996) Can destination management systems provide competitive advantage? A discussion of the factors affecting the survival and success of destination management systems. *Progress in Tourism and Hospitality Research* 2, 1–13.

Beeton, S. (2005) In: Brent Ritchie, J.R. and Crouch, G.I. (eds) *The Competitive Destination: A Sustainable Tourism Perspective.* CAB International, Wallingford, UK, pp. 294–296.

Benito, B., Solana, J. and López, P. (2014) Determinants of Spanish regions' tourism performance: A two-Stage, double-bootstrap data envelopment analysis. *Tourism Economics* 20, 987–1012.

Berger, T. (2008) Concepts on national competitiveness. *Journal of International Business and Economy* 9, 3–17.

Blake, A., Sinclair, M.T. and Soria, J.A.C. (2006) Tourism productivity: evidence from the United Kingdom. *Annals of Tourism Research* 33, 1099–1120.

Bristow, G. (2005) Everyone's a 'winner': Problematising the discourse of regional competitiveness. *Journal of Economic Geography* 5, 285–304.

Buhalis, D. (2000) Marketing the competitive destination of the future. *Tourism Management* 21, 97–116.

Campos-Soria, J.A., González García, L. and Ropero García, M.A. (2005) Service quality and competitiveness in the hospitality sector. *Tourism Economics* 11, 85–102.

Carmichael, B.A. (2002) Global competitiveness and special events in cultural tourism: the example of the Barnes Exhibit at the Art Gallery of Ontario, Toronto. *CAG Canadian Geographer* 46, 310–324.

Chacko, H.E. (1996) Positioning a tourism destination to gain a competitive edge. *Asia Pacific Journal of Tourism Research* 1, 69–75.

Chen, C.-M., Chen, S.H. and Lee, H.T. (2011) The destination competitiveness of Kinmen's tourism industry: Exploring the interrelationships between tourist perceptions, service performance, customer satisfaction and sustainable tourism. *Journal of Sustainable Tourism* 19, 247–264.

Chens, C.-Y., Sok, P. and Sok, K. (2008) Evaluating the competitiveness of the tourism industry in Cambodia: Self-assessment from professionals. *Asia Pacific Journal of Tourism Research* 13, 41–66.

Cho, D.-S. and Moon, H.C. (1998) A nation's international competitiveness in different stages of economic development. *Journal of Competitiveness Studies* 6, 5.

Chon, K.-S. and Mayer, K.J. (1995) Destination competitiveness models in tourism and their application to Las Vegas. *Journal of Tourism Systems and Quality Management* 1, 3.

Chon, K.-S., Weaver, P.A. and Kim, C.-Y. (1991) Marketing your commmunity: image analysis in Norfolk. *Cornell Hotel and Restaurant Administration Quarterly* 31, 31–37.

Craigwell, R. (2007) *Tourism Competitiveness in Small Island Developing States*. UNU World Institute for Development Economics Research, Helsinki, Finland.

Croes, R. (2010) Small island tourism competitiveness: expanding your destination's slice of paradise. *Rosen College of Hospitality Management, Lecture Notes*, 1–32.

Croes, R. (2011) Measuring and explaining competitiveness in the context of small island destinations. *Journal of Travel Research* 50, 431–442.

Croes, R. (2012) Assessing tourism development from Sen's capability approach. *Journal of Travel Research* 51, 542–554.

Croes, R. and Kubickova, M. (2013) From potential to ability to compete: towards a performance-based tourism competitiveness index. *Journal of Destination Marketing and Management* 2, 146–154.

Croes, R. and Rivera, M.A. (2010) Testing the empirical link between tourism and competitiveness: evidence from Puerto Rico. *Tourism Economics* 16, 217–234.

Crouch, G.I. (2011) Destination competitiveness: an analysis of determinant attributes. *Journal of Travel Research* 50, 27–45.

Crouch, G.I. and Ritchie, J.R.B. (1999) Tourism, competitiveness, and societal prosperity. *Journal of Business Research* 44, 137–152.

Crouch, G.I. and Ritchie, J.B. (2000) The competitive destination: a sustainability perspective. *Tourism Management* 21, 1–7.

Cunha, S.K.D. and Cunha, J.C.D. (2005) Tourism cluster competitiveness and sustainability: proposal for a systemic model to measure the impact of tourism on local development. *BAR – Brazilian Administration Review* 2, 47–62.

Das, J. and DiRienzo, C. (2010) Tourism competitiveness and corruption: A cross-country analysis. *Tourism Economics* 16, 477–492.

De Keyser, R. and Vanhove, N. (1994) The competitive situation of tourism in the Caribbean area methodological approach. *The Tourist Review* 49, 19–22.

D'Hauteserre, A.-M. (2000) Lessons in managed destination competitiveness: The case of Foxwoods Casino Resort. *Tourism Management* 21, 23–32.

Dwyer, L. and Forsyth, P. (2011) Methods of estimating destination price competitiveness: A case of horses for courses? *Current Issues in Tourism* 14, 751–777.

Dwyer, L. and Kim, C. (2003) Destination competitiveness: Determinants and indicators. *Current Issues in Tourism* 6, 369–414.

Dwyer, L., Forsyth, P., Madden, J. and Spurr, R. (2000a) Economic impacts of inbound tourism under different assumptions regarding the macroeconomy. *Current Issues in Tourism* 3, 325–363.

Dwyer, L., Forsyth, P. and Rao, P. (2000b) The price competitiveness of travel and tourism: A comparison of 19 destinations. *Tourism Management* 21, 9–22.

Dwyer, L., Mistilis, N., Forsyth, P. and Rao, P. (2001) International price competitiveness of Australia's mice industry. *International Journal of Tourism Research* 3, 123.

Dwyer, L., Forsyth, P. and Rao, P. (2002) Destination price competitiveness: Exchange rate changes versus domestic inflation. *Journal of Travel Research* 40, 328–336.

Dwyer, L., Livaic, Z. and Mellor, R. (2003) Competitiveness of Australia as a tourist destination. *Journal of Hospitality and Tourism Management* 10(1), 60–78.

Dwyer, L., Mellor, R., Livaic, Z., Edwards, D. and Kim, C. (2004) Attributes of destination Competitiveness: A factor analysis. *Tourism Analysis* 9, 91–101.

Enright, M.J. and Newton, J. (2004) Tourism destination competitiveness: A quantitative approach. *Tourism Management* 25, 777–788.

Enright, M.J. and Newton, J. (2005) Determinants of tourism destination competitiveness in Asia Pacific: Comprehensiveness and universality. *Journal of Travel Research* 43, 339–350.

Evans, N., Campbell, D. and Stonehouse, G. (2002) *Strategic Management for Travel and Tourism.* Butterworth-Heinemann, Oxford, UK.

Fagerberg, J. (1996) Technology and Competitiveness. *Oxford Review of Economic Policy* 12(3), 39–51.

Faulkner, B., Oppermann, M. and Fredline, E. (1999) Destination competitiveness: An exploratory examination of South Australia's core attractions. *Journal of Vacation Marketing* 5, 125–139.

Go, F.M. and Govers, R. (2000) Integrated quality management for tourist destinations: A European perspective on achieving competitiveness. *Tourism Management* 21, 79–88.

Gomezelj Omerzel, D. (2011) The local business sector's perception of the competitiveness of Slovenia as a tourist destination. *Turizam: znanstveno-stručni časopis* 59, 25–46.

Gooroochurn, N. and Sugiyarto, G. (2005) Competitiveness indicators in the travel and tourism industry. *Tourism Economics* 11, 25–43.

Guo, S. and Cao, N. (2004) An analysis of urban tourism competition. *Commercial Research* 9, 138–140.

Gursoy, D., Baloglu, S. and Chi, C.G. (2009) Destination competitiveness of Middle Eastern countries: An examination of relative positioning. *Anatolia: An International Journal of Tourism and Hospitality Research* 20, 151–163.

Hassan, S.S. (2000) Determinants of market competitiveness in an environmentally sustainable tourism industry. *Journal of Travel Research* 38, 239–245.

Heath, E. (2003) Towards a model to enhance destination competitiveness: A Southern African perspective. *Proceeding of the Council for Australian University Tourism and Hospitality Education (CAUTHE) 2003 National Research Conference*, 2003, Coffs Harbour, Australia.

Hong, W.-C. (2009) Global competitiveness measurement for the tourism sector. *Current Issues in Tourism* 12, 105–132.

Hu, Y. and Ritchie, J.R.B. (1993) Measuring destination attractiveness: a contextual approach. *Journal of Travel Research* 32, 25–34.

Hudson, S., Ritchie, B. and Timur, S. (2004) Measuring destination competitiveness: An empirical study of Canadian ski resorts. *Tourism and Hospitality Planning and Development* 1, 79–94.

Huybers, T. and Bennett, J. (2003) Inter-firm cooperation at nature-based tourism destinations. *Journal of Socio-Economics* 32, 571–587.

Icoz, O., Var, T. and Kozak, M. (1998) Tourism demand in Turkey. *Annals of Tourism Research* 25, 236–240.

IMD (2013) *IMD World Competitiveness Yearbook* [Online]. Available: http://www.imd.org/news/World-Competitiveness-2013.cfm (accessed 17 January 2014).

Jackman, M., Lorde, T., Lowe, S. and Alleyne, A. (2011) Evaluating tourism competitiveness of small island developing states: A revealed comparative advantage approach. *Anatolia* 22, 350–360.
Jamal, T.B. and Getz, D. (1996) Does strategic planning pay? Lessons for destinations from corporate planning experience. *Progress in Tourism and Hospitality Research* 2, 59–78.
Kayar, Ç.H. and Kozak, N. (2010) Measuring destination competitiveness: An application of the travel and tourism competitiveness Index (2007). *Journal of Hospitality Marketing and Management* 19, 203–216.
Kaynak, E. and Marandu, E.E. (2006) Tourism market potential analysis in Botswana: A Delphi study. *Journal of Travel Research* 45, 227–237.
Kim, S.-S., Crompton, J.L. and Botha, C. (2000) Responding to competition: A strategy for Sun/Lost City, South Africa. *Tourism Management* 21, 33–41.
Kim, S.S., Yoon, S. and Kim, Y. (2011) Competitive positioning among international convention cities in the East Asian region. *Journal of Convention and Event Tourism* 12, 86–105.
Kozak, M. (2003) Measuring tourist satisfaction with multiple destination attributes. *Tourism Analysis* 7, 229–240.
Kozak, M. and Rimmington, M. (1999) Measuring tourist destination competitiveness: conceptual considerations and empirical findings. *International Journal of Hospitality Management* 18, 273–283.
Krugman, P. (1994) Competitiveness: A dangerous obsession. *Foreign Affairs* 73, 2.
Krugman, P.R. (1996) Making sense of the competitiveness debate. *Oxford Review of Economic Policy* 12, 17–25.
Lall, S. (2001) Competitiveness indices and developing countries: An economic evaluation of the global competitiveness report. *World Development* 29, 1501–1525.
Lee, Y. and Chen, C. (2010) Examining and comparing the competitiveness of tourism industry in Cambodia and Taiwan: An assessment from professionals. *Journal of American Academy of Business* 16, 129–140.
Leiper, N. (2008) Why 'the tourism industry' is misleading as a generic expression: The case for the plural variation, 'tourism industries'. *Tourism Management* 29, 237–251.
Lew, A. and McKercher, B. (2006) Modeling tourist movements: A local destination analysis. *Annals of Tourism Research* 33, 403–423.
Li, G., Song, H., Cao, Z. and Wu, D.C. (2013) How competitive is Hong Kong against its competitors? an econometric study. *Tourism Management* 36, 247–256.
Li, W., Yang, Y. and Zeng, K. (2007) Tourism destination system and core competitiveness. *An Hui Agriculture Science* 35, 8320–8321.
Mangion, M.-L., Durbarry, R. and Sinclair, M.T. (2005) Tourism competitiveness: Price and quality. *Tourism Economics* 11, 45–68.
Mathew, V. (2009) Sustainable tourism: A case of destination competitiveness in South Asia. *South Asian Journal of Tourism and Heritage* 2, 83–89.
Mazanec, J.A. (1995) Competition among European tourist cities: A comparative analysis with multidimensional scaling and self-organizing maps. *Tourism Economics* 1, 283–302.
Mazanec, J.A. and Ring, A. (2011) Tourism destination competitiveness: Second thoughts on the World Economic Forum reports. *Tourism Economics* 17, 725–751.
Mazanec, J.A., Wöber, K. and Zins, A.H. (2007) Tourism destination competitiveness: From definition to explanation? *Journal of Travel Research* 46, 86–95.
Mihalič, T. (2000) Environmental management of a tourist destination: A factor of tourism competitiveness. *Tourism Management* 21, 65–78.
Miller, M.M., Henthorne, T.L. and George, B.P. (2008) The competitiveness of the Cuban tourism industry in the twenty-first century: A strategic re-evaluation. *Journal of Travel Research* 46, 268–278.
Mingaleva, Z. and Bunakov, O. (2014) Competitiveness assessment of Russian territories in terms of inbound tourism. *Life Science Journal* 11, 318–321.

OECD (1994) *Technology and the Economy: The Key Relationships.* Organization for Economic Cooperation and Development, Paris.

Papatheodorou, A. (2002) Exploring competitiveness in Mediterranean resorts. *Tourism Economics* 8, 133–150.

Papatheodorou, A. and Song, H. (2005) International tourism forecasts: Time-series analysis of world and regional data. *Tourism Economics* 11, 11–23.

Pearce, D.G. (1997) Competitive destination analysis in Southeast Asia. *Journal of Travel Research* 35, 16–24.

Pike, S. (2005) Tourism destination branding complexity. *Journal of Product and Brand Management* 14, 258–259.

Porter, M.E. (1985) *Competitive Advantage: Creating and Sustaining Superior Performance.* Free Press, New York.

Porter, M.E. (1990) *The Competitive Advantage of Nations.* Free Press, New York.

Ritchie, J.R.B. and Crouch, G.I. (1993) *Competitiveness In International Tourism: A Framework for Understanding and Analysis.* World Tourism Education and Research Centre, University of Calgary, Calgary, Canada.

Ritchie, J.R.B. and Crouch, G.I. (2003) *The Competitive Destination: a Sustainable Tourism Perspective.* CAB International, Wallingford, UK.

Saayman, A. and Saayman, M. (2008) Determinants of inbound tourism to South Africa. *Tourism Economics* 14, 81–96.

Sahli, M. (2006) Tourism destination specialization. In: Dwyer, L. and Forsyth, P. (eds) *International Handbook on the Economics of Tourism.* Edward Elgar, Cheltenham, UK.

Schalber, C. and Peters, M. (2012) Determinants of health tourism competitiveness: An Alpine case study. *Turizam: znanstveno-stručni časopis* 60, 307–323.

Sen, A. (1999) *Development as Freedom* Knopf, New York.

Smit, A. (2010) The competitive advantage of nations: Is Porter's diamond framework a new theory that explains the international competitiveness of countries? *Southern African Business Review* 14(1).

Song, Z. (2008) On the cultivation of tourism destination key competitiveness. *Reformation and Strategy* 24, 95–99.

Soteriou, E.C. and Roberts, C. (1998) The strategic planning process in national tourism organizations. *Journal of Travel Research* 37, 21–29.

Stevens, B.F. (1992) Price value perceptions of travelers. *Journal of Travel Research* 31, 44–48.

Vlami, A., Zacharatos, G.A. and Tsamos, G. (2006) Government involvement in regionalization of tourism development in Greece. *Proceedings of the 2nd International Conference on Geotour,* Kosice, Slovakia, pp. 5–16.

Wang, C.-Y., Hsu, M.K. and Swanson, S.R. (2012) Determinants of tourism destination competitiveness in China. *Journal of China Tourism Research* 8, 97–116.

WEF (2013) Travel and tourism competitiveness report [Online]. Available at: https://www.weforum.org/reports (accessed 24 March 2014).

Wint, A.G. (2003) *Competitiveness in Small Developing Economies Insights from the Caribbean.* University of the West Indies Press, Barbados.

World Economic Forum (2015) *The Global Competitiveness Report* [Online]. World Economic Forum. Available at: http://www.weforum.org (accessed 2 February 2016).

Zang, D.-X. and Huang, J. (2007) An empirical analysis on influencing factors of seaside tourism destination competitiveness. *Geography and Geo-Information Science* 5, 24.

5 Island Tourism Competitiveness and Sustainability in the Maldives

Jack Carlsen* and Mariyam Zulfa

Curtin University, Western Australia

5.1 Introduction

This chapter will investigate the evolution of destination competitiveness factors in the Maldives tourism sector in an era when market sensitivity and volatility have increased in response to economic, political, social and environmental changes (Ritchie and Crouch, 2003) and new sources of competition are emerging. For island nations such as the Maldives, the natural features of the destination are now just as important as the built facilities in providing a unique selling point and source of competitive advantage in the holiday travel market. Islands that have sustainably managed their marine and terrestrial environments can enjoy a comparative advantage in being able to transform these physical factors of production (Porter, 1990) into economic advantage through tourism.

Maldives is one such destination that has successfully exploited its natural assets to establish a thriving and competitive tourism sector. The volume of annual international tourist arrivals has risen from around 42,000 in 1980 to around 460,000 in 2000, a more than ten-fold increase over that 20-year growth period. By 2010, annual international tourist arrivals reached 790,000, after several years of declining visitor arrivals in 2005 and 2009 due to the Indian Ocean tsunami and the global financial crisis, respectively (MOTAC, 2011).

The difficult market conditions over the last 5 years have increased the focus on re-establishing the growth trend in arrivals that the Maldives had enjoyed since the inception of tourism in the 1980s. This was made more difficult in an environment of increasing competition amongst island tourism destinations and a static or declining number of tourists worldwide due to the combined effects of economic uncertainty and global insecurity (MOTAC, 2013). Therefore it became necessary for the Maldives to understand more deeply the factors that provide the sources of competitive advantage and take action to capitalize on those.

*E-mail: jackcarlsen2@bigpond.com

Whilst some sources of competitive advantage may be immediately apparent (good weather, clean waters), it is necessary to move beyond these physical factors in order to identify the true nature of competitive advantage in tourism destinations such as the Maldives. In this way, tourism planning, marketing and management policies and programmes can be targeted towards maintaining competitive advantage and translating it into increased tourist arrivals over time.

This chapter explores the factors that constitute destination competitiveness, first by reviewing previous studies and synthesizing that literature into a matrix of variables, survey instruments and units of analysis previously employed to measure competitiveness. From this synthesis, six groups comprising 48 competitiveness factors were tested on a range of tourism experts in the Maldives to assess both the importance and performance of each variable. Importance-Performance Analysis (IPA) grids (Oh, 2001) provide a summary of findings and a means of identifying and assessing the factors that make the Maldives competitive and that therefore should provide the focus of sustainable competitive advantage. IPA also identifies areas where the Maldives could improve its competitive performance and these should be priority areas that are addressed in future policy, planning and management activities.

5.2 Literature Review

The study of tourism's competitiveness had its beginnings in the examination of organizational level strategies, conducted with a view to establishing a nexus between industry practice and academic research (Poon, 1993; Crouch and Ritchie, 1999; Buhalis, 2000). One of the earliest published researches into organizational strategy in the tourism industry was conducted by Poon (1993), whose study involved documenting insights into marketing practices made in some destination nations in the Caribbean.

Poon theorized that the disparate social groupings and varying ages of the travelling public must be taken into account when creating competitive strategy and argued for a varied range of approaches to destination-marketing, rather than employing the mass-marketing methods that were in vogue at the time (Poon, 1993). Poon contributed to a change in strategic thinking amongst tourism researchers in how marketers should perceive a destination: it is an amalgam of a broad range of products suited to a diversified range of market segments; therefore, strategy accordingly must reflect that reality. Predominant views of mass-tourism or 'Fordian' forms of tourism (Poon, 1993) which influenced strategic thinking in tourism marketing in the mid-1980s changed gradually as a result of these conceptualizations.

Another predominant trend of the 1980s was the amalgamation of major airlines to cater for increases in traffic flows. Poon theorized that the occurrence of such trends was the result of firms adopting flexible approaches to specialization. While such theory-making was not backed by adequate supporting data empirically, her observations contributed to rethinking existing practices in destination marketing. A call was made within tourism research for new ways to provide more innovative and market-oriented products and services within the tourism destination. Poon's model of flexible-specialization enabled the adopting of innovative approaches to strategic marketing and improving destination competitiveness (Poon, 1993).

Another important study within tourism research which focused sharply on *sustainable development* is Hassan's (2000) work, which provided a model that destinations must adopt if sustainable competitiveness was to be achieved. In this way, Hassan defines competitiveness as 'a destination's ability to create and integrate value-added products that sustain its resources while maintaining market position relative to competitors' (Hassan, 2000: 240). Hassan largely based his model on observed trends, not on empirical findings, and his conviction that 'sustainable development has taken centre stage in strategic planning' (p. 242).

Hassan observes that market segments were rapidly 'greening'; subsequently, the model was based on his own observations derived from the literature that supported the rising trend of 'green' tourism. It followed for Hassan that any market-driven strategy must be based on the needs of populations of travellers who demanded sustainable forms of tourism. In Hassan's view, sustainability is achieved through commitment to environmental preservation. This should be the predominant goal of any strategy that vies for competitiveness, and any model of competitiveness must centre on sustainability; hence one facet of his model is dedicated to the destination's environmental commitment.

The most important contribution to sustainable destination competitiveness has come from Ritchie and Crouch (2000, 2003). They used a combination of resource-based and strategic factors to analyse competitiveness of tourism. Crouch and Ritchie (2003) regard a destination as competitive if it is able to increase expenditure by tourists within the destination, and to attract visitors while providing them with satisfying experiences. In their view it must do so in a profitable way, while enhancing the *well-being* of the population of the destination and at the same time preserving its natural endowments (Ritchie and Crouch, 2003: 2). Their model of competitiveness, which is the most widely cited in the tourism literature, offered a variety of factors as impacting on competitiveness.

Competitiveness on the basis of price or price-competitiveness of the tourism industry has been studied extensively by Dwyer and Kim (2003). In that context, competitiveness is defined by these authors as a general concept that encompasses 'price differentials coupled with exchange rate movements, productivity levels of various components of the tourist industry and qualitative factors affecting the attractiveness or otherwise of a destination' (p. 374). Price competitiveness within tourism research is generally considered a part of the broader concept of economic competitiveness (Ritchie and Crouch, 2003). If analysed from this viewpoint, all destinations will either vie to offer lower cost products than their competitors or compete on an advantage or advantages derived from unique features; not unlike the 'diamond' model of competitive-advantage based on the philosophy of Porter (1990). In that way, by offering competitively priced products or those that are unique, destinations are increasing their income, thereby increasing their societal well-being (Crouch and Ritchie, 1999).

Tourism destinations exist at a number of different geographical levels that are inter-related (Swarbrooke, 2002), so when making comparisons it is logical to assume that only those with a related set of attributes can offer meaningful measures and homogeneity of variance (Field, 2005). Enright and Newton (2005) specify that in measuring competitiveness 'the same determinants [must] have relative importance in [the] destinations [compared]' and they postulate that, to avoid 'incorrect comparisons', the destinations or units of analysis selected must be relatively comparable in terms of destination-type (Enright and Newton, 2005); for example, urban tourism destinations may only be compared with other similar urban destinations. Theoretically, incorrect comparisons can thus lead to 'confusion' in their non-conformity to homogeneity (Kline, 2000).

Table 5.1. Matrix of factors engaged by relevant contemporary research projects including this research.

Enright *et al.* (1997)	**Crouch and Ritchie (1999)**	**Dwyer and Kim (2003)**	**Gooroochurn and Sugiyarto (2005)**	**Enright and Newton (2004, 2005)**	**Zulfa (2009)**
Variables: 1. Inputs 2. Industrial & Regional Clustering 3. Interfirm Competition and Co-operation 4. Industrial and Consumer Demand 5. Institutions and Agendas 6. Internal Organization and Strategy of firms	**Variables:** 1. Qualifying & Amplifying Determinants 2. Destination policy, planning and development 3. Destination management 4. Core Resources and Attractors 5. Supporting Factors and Resources	**Variables:** 1. Inherited Resources (Natural resources) 2. Created Resources (Tourism infrastructure Special events Activities Entertainment Shopping) 3. Supporting factors and Resources (General infrastructure, accessibility, quality, market ties, Hospitality) 4. Destination Management 5. Demand Conditions 6. Situational Conditions	**Variables:** 1. Price 2. Openness 3. Technology 4. Infrastructure 5. Human tourism 6. Social development 7. Environment 8. Human resources	**Variables:** 1. Cuisine 2. Safety 3. Nightlife 4. Visual appeal 5. Climate 6. Well-known landmarks 7. Different culture 8. Local way of life 9. Special events 10. Interesting architecture 11. Interesting festivals 12. Dedicated tourism attractions 13. Notable history 14. Music and performances 15. Museums and galleries	**Independent variables:** 1. Physical Factors 2. Man-made Factors 3. Related and Supporting Factors 4. Price Factors 5. Destination Management Factors 6. Other factors (relating to social, legal, political and environmental contexts)
Surveys: Industry	**Surveys:** None	**Surveys:** Product and service providers	**Surveys:** None: Secondary data only	**Surveys:** Tourism industry service-providers	**Surveys:** Tourism industry service providers and destination experts
Unit of analysis: Industry (Hong Kong)	**No units observed**	**Unit of analysis:** Tourism industry, Australia and Korea	**Unit of analysis**: The world system	**Unit of analysis:** Tourism Industry, Singapore, Hong Kong, Thailand	**Unit of analysis:** Small-islands/marine tourism destinations

Table 5.1 lists variables used in relevant contemporary research into tourism destination competitiveness. As evident, there is now a greater degree of consensus on the content-matter of attributes chosen in research designs for the study of competitiveness. Generally, *physical and man-made factors* peculiar to destinations were regarded as key sources of destination attractiveness (Crouch, 2006). These factors constituted the main reason why tourists would visit a particular destination. Then the *destination management factors* supplemented these, and together they assisted in creating the context in which competitive advantages arose for each destination, that is to say a context in which *physical and man-made factors* were strategically exploited to devise ways to render the destination more attractive to visit.

5.3 Method

The research design employed in this study involved both qualitative and quantitative methodologies; qualitative in the sense that respondents were asked to rate their opinion; quantitative methods were then deployed to scale the responses derived from the survey results by applying the scaling technique (Jick, 1979). Quantitative methodologies were employed in investigating the correlation amongst variables and attributes and these were used supplementarily to support or discard findings from the survey.

The attributes used in the survey were derived without bias from the literature, and testing by means of a focus group was used to ensure that respondents understood the meaning of terms employed, prior to the survey's distribution, which was conducted over a 6-month duration. Attributes that were salient to the competitiveness of island tourism destinations were chosen to be included in the survey.

Enright and Newton (2005) as well as Dwyer and Kim (2003) employed survey methods in which respondents were asked to rate the importance of attributes as they were perceived by respondents, as contributing to tourism's competitiveness. Their respective groups of respondents comprised of members of tourism associations. A similar ratings method was used by Crouch (2006) when he opened an online survey based on the Ritchie and Crouch model, targeted at tourism researchers, experts and industry participants. Tourists themselves were not surveyed in either of these research projects. Consistent with those methods, the survey for this research asked respondents to rate the importance of attributes that denoted competitiveness as perceived by them. Following these researchers, members of Destination Management Organizations (DMOs) and tourism destination experts are designated as the appropriate representatives of the unit of analysis. The antecedent factors of competitiveness for small islands/marine tourism destinations were derived from the extant literature.

The survey constituted six groups comprising 48 competitiveness factors and asked respondents to rate their importance to competitiveness as well as the performance in the Maldives. The six factor groupings and related competitiveness factors were:

1. Physical Factors (7)
    1. Comfortable climate and weather conditions
    2. Clean waters (lagoons and surrounding seas)
    3. Marine-based flora and fauna

4. Beaches
5. Land flora and fauna
6. Scenery
7. Conditions for water sports

2. Man-made Factors (11)
8. Scuba diving facilities
9. Range of water activities
10. Welcoming host community
11. Fishing opportunities
12. Local cultural events
13. International fairs and events
14. Artefacts and monuments of historical significance
15. Museums and art galleries
16. Night-life
17. Range of shopping outlets
18. Business and government functions

3. Related and Supporting Factors (12)
19. Skilled and friendly staff
20. International airlines
21. Sanitation and hygiene
22. Availability of phone/fax/internet
23. Range of accommodation
24. Range of activities
25. Range of food and beverage outlets
26. Promotes foreign investment
27. Promotes local investment
28. Helpful airport staff
29. Health and medical facilities
30. Airports' infrastructure

4. Price Factors (4)
31. Prices of air tickets
32. Value-for-money accommodation
33. Prices of activities
34. Transport prices in destination

5. Destination Management Factors (10)
35. DMO involvement in marketing
36. Resources for marketing
37. Environmental protection
38. Preparedness for crises
39. Cooperation in image-control
40. Public–private sector cooperation
41. Financial support for marketing
42. Local community involvement
43. Policies that support profitability
44. Hospitality training opportunities

6. Other Factors (4)
   45. Safety and security
   46. Political stability in destination
   47. Religious and social harmony
   48. Business-friendly legal and regulatory environment

Importance-Performance analysis (IPA) (Crompton and Duray, 1985) was employed to frame the findings, and is a three-step process, which involves:

1. The identification of key attributes through a literature review, focus group interviews or other techniques.
2. Surveying of respondents to determine the importance and performance of each attribute.
3. Calculating importance and performance scores for each attribute and presenting these on a two-dimensional IPA grid.

The IPA grid is divided into four quadrants based on the importance and performance scores of the attributes. The four quadrants are labelled: 'Concentrate Here' (high importance, low performance), 'Keep Up the Good Work' (high importance, high performance), 'Low Priority' (low importance, low performance) and 'Possible Overkill' (low importance, high performance).

While the most common applications of IPA are found in the marketing field, the use of IPA has been extended beyond consumer evaluations to include other types of respondents. Evans and Chon (1989) used the approach to survey business managers to develop and evaluate tourism policy in two separate mountain destinations in the USA. The analysis presented in this article uses a similar approach by combining IPA with an expert informant approach (Liburd *et al.*, 2011).

Multi-dimensional scaling was employed to present the analysis results in four importance and performance quadrants. Quadrant one includes factors that scored high in both performance and importance and those attributes that destinations needed to 'keep up the good work' (Martilla and James, 1977). Quadrant two contains factors that scored low in importance but high on performance and these attributes require no further attention or resources. Quadrant three has factors that are low in importance as well as performance, indicating that DMOs need not be concerned about factors in this quadrant. Most importantly, quadrant four comprises factors that are high in importance but scored low in performance, hence identifying areas that DMOs need to pay special attention to in order to enhance competitiveness in the destination.

The specific factors were derived from the literature review and were considered as significant to the competitiveness of small islands/marine tourism destinations. Respondents also had the opportunity to add any other comments that they considered relevant to competitiveness of the Maldives.

5.3.1 Sampling

A total of 46 respondents from the Maldives, of which close to 50% were in the category Chairpersons, Chief Executive Officers and Managing Directors and 36% were in the General Managers, Managers and Deputies category and hence highly representative of DMO activities. Despite some limitations in sample size and representativeness, the depth and quality of the responses received were significant. Over 80% of the respondents themselves were in key positions in DMOs and the experts represented a cross-section of key academic experts in the field in tourism research and hence the sample comprises a comprehensive representation of key industry stakeholder and expert opinion. The survey responses have satisfied the validity criteria in that the application of findings can be generalized across tourism destinations. More or less responses would not have altered the key findings and the construct validity of the research attributes has been confirmed within the sample used for investigative purposes.

5.4 Analysis

5.4.1 Physical factors

Analysis of survey data on *Physical Factors* showed that the Maldives was performing well in the following important competitiveness factors (see quadrant one, Fig. 5.1).

1. Comfortable climate and weather conditions
2. Clean waters (lagoons and surrounding seas)
3. Marine-based flora and fauna
4. Beaches

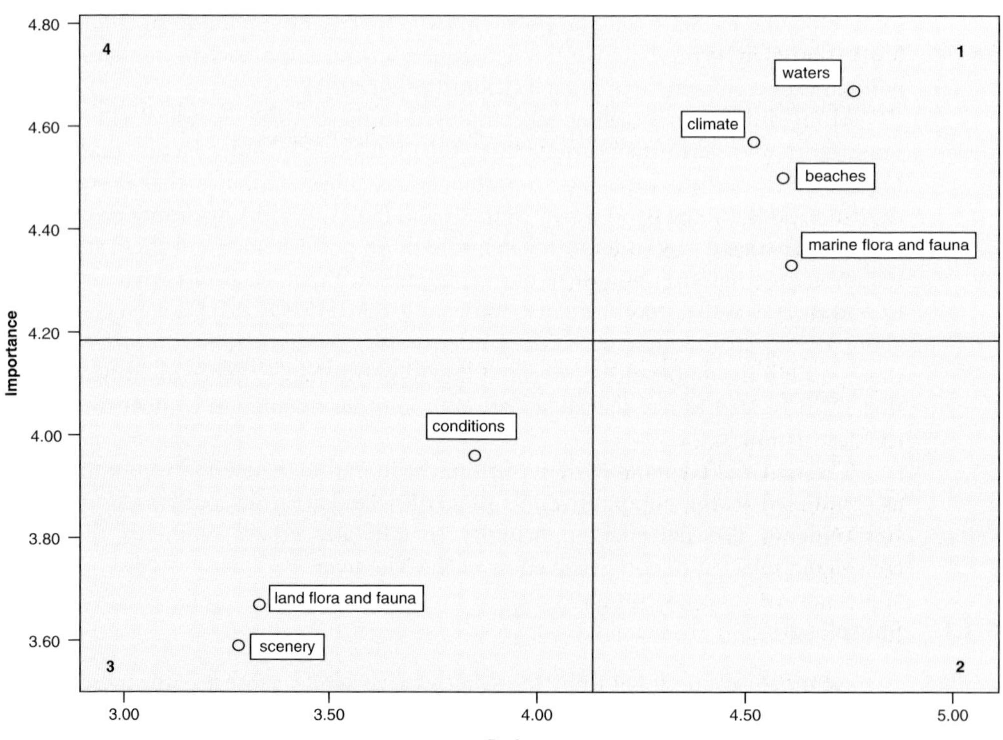

Fig. 5.1. Physical factors.

The following factors were considered less important and less well-performed in the Maldives (see quadrant three, Fig. 5.1):

5. Land flora and fauna
6. Scenery
7. Conditions for water sports

The following additional comments demonstrate the importance of physical and environmental conditions in the Maldives in creating sustainable competitive advantage:

- Good weather, pristine beaches and a good product designed to customer perception, blended with the ingredient of privacy makes a destination very competitive;
- Sustainable management of physical attractions is important. The integration of physical development with the natural setting is vital to be competitive;
- Scuba diving is very important. Also architectural concepts used in the design of the resort, upgrading, friendly people and atmosphere all made Maldives a success;
- Salubrious, warm climate is most important. Sun lust is still the main driver for tourists from the northern hemisphere;
- With weather conditions and patterns changing all over the world, tourism operators of the world need better prediction of weather conditions. More research on changing ecology is important to every destination.

5.4.2 Man-made factors

Analysis of survey data on *Man-made Factors* showed that the Maldives was performing well in the following important competitiveness factors (see quadrant one, Fig. 5.2):

8. Scuba-diving facilities
9. Range of water activities
10. Welcoming host community
11. Fishing opportunities

The following factors were considered less important and less well-performed in the Maldives (see quadrant three, Fig. 5.2):

12. Local cultural events
13. International fairs and events
14. Artefacts and monuments of historical significance
15. Museums and art galleries
16. Night-life
17. Range of shopping outlets
18. Business and government functions

Hence it can be stated that the results strongly indicate that man-made factors were not important for competitiveness, with the exception of the factors relating to water-based activities and having a welcoming host community which enabled the Maldives to be competitive.

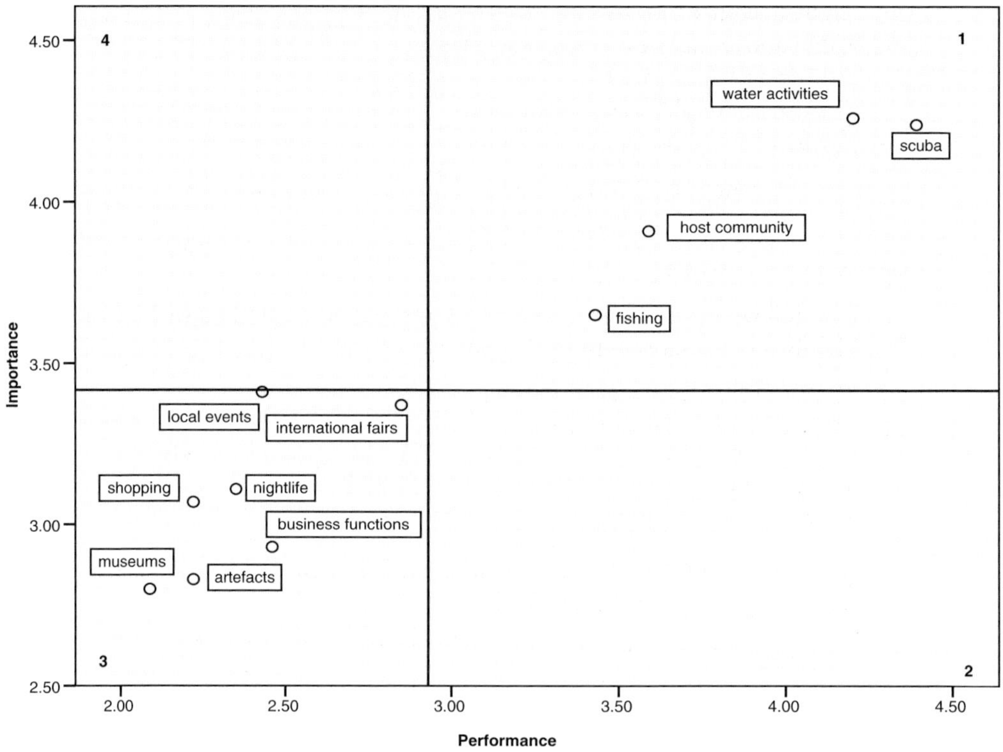

Fig. 5.2. Man-made factors.

In addition, the following comments were returned by respondents on *Man-made factors*, illuminating the above responses further:

- More than just the physical characteristics of the destination, attitude and customer-friendly community in the destination is extremely critical for a destination being competitive;
- When it comes to small islands destinations no tradition, man-made activity or entertainment level can be considered a key factor of competitiveness as none of them could ever create a stronger demand as the destination itself does. The basis for this is the fact that travellers to small islands destinations are 99% leisure tourists seeking exactly what the specific destination has to offer. No single tourist comes here seeking activity, entertainment, shopping, etc. on a grand scale, at least not at the current time. Surely the history of tourism in Maldives is based on diving, etc. and there is still a fairly large amount of activity-based all-inclusive resorts (usually catering to an all-Italian clientele). However, in my opinion they do not contribute to the competitiveness of the country/destination as a whole.

5.4.3 Related and supporting factors

Analysis of survey data on *Related and Supporting Factors* showed that the Maldives was performing well in the following important competitiveness factors (see quadrant one, Fig. 5.3):

19. Skilled and friendly staff
20. International airlines
21. Sanitation and hygiene
22. Availability of phone/fax/internet
23. Range of accommodation

The following factor was considered less important but well-performed in the Maldives (see quadrant two, Fig. 5.3):

24. Range of activities

The following factors were considered less important and less well-performed in the Maldives (see quadrant three, Fig. 5.3):

25. Range of food and beverage outlets
26. Promotes foreign investment
27. Promotes local investment
28. Helpful airport staff
29. Health and medical facilities

The following factor was considered more important and less well-performed in the Maldives (see quadrant four, Fig. 5.3):

30. Airports' infrastructure

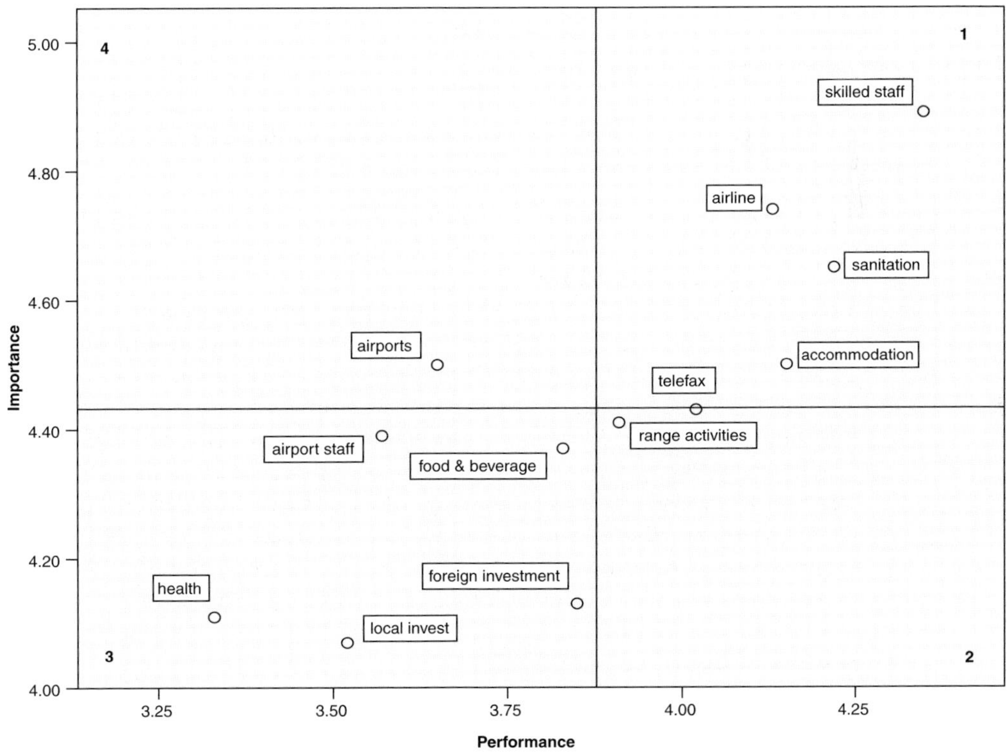

Fig. 5.3. Related and supporting factors.

The following comments were returned by respondents:

- The number of airlines coming into Maldives is limiting our potential;
- With regard to investments in Maldives, what attracted investors was the lack of corporate tax and income tax. I am not sure about the facilities provided to encourage investments. Lack of capital financing sources, high interest rates from the banks and the current bidding system (an advance payment of 5 million dollars) definitely do not encourage investments;
- Health and medical facilities are not important for a visitor. But it must be available if required. Travel assistance (insurance) is now included in most tour operator's packages and it is used very well if required.

5.4.4 Price factors

Analysis of survey data on *Price Factors* showed that the Maldives was performing well in the following important competitiveness factors (see quadrant one, Fig. 5.4):

31. Prices of air-tickets
32. Value-for-money accommodation

The following factors were considered less important and less well-performed in the Maldives (see quadrant three, Fig. 5.4):

33. Prices of activities
34. Transport prices in destination

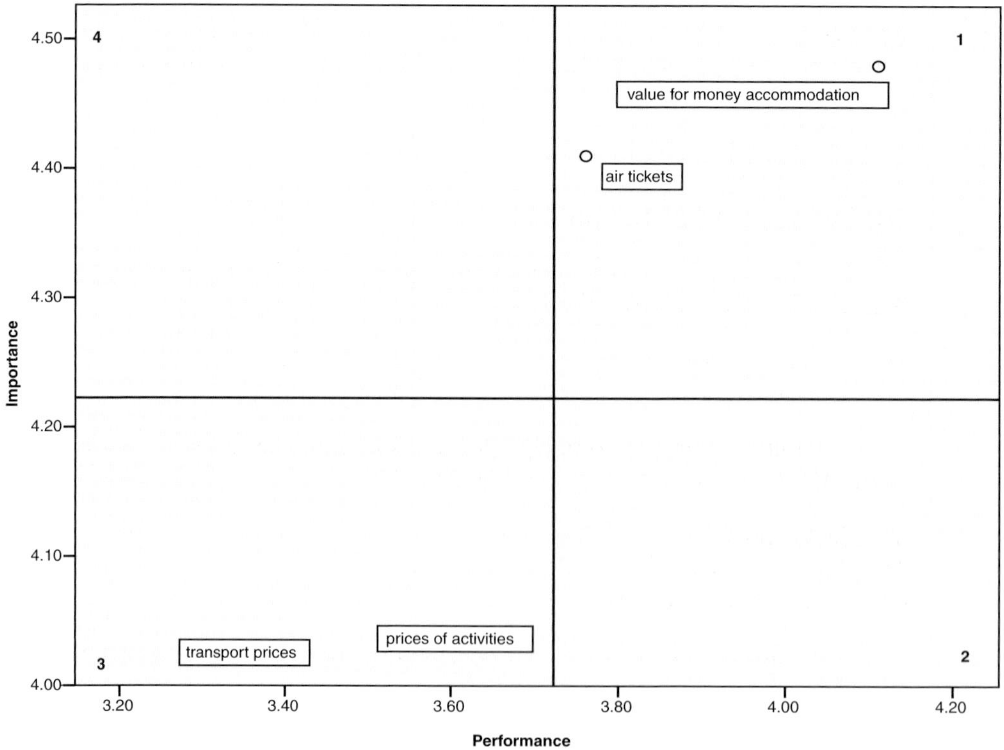

Fig. 5.4. Price factors.

The following comments were returned by respondents:

- Prices of all products and services play a critical role;
- Clients do check and compare prices before they buy. However, if they select a particular place they just take it rather than being price-conscious. Price is not a matter where luxury is concerned;
- Service and value for money are key factors – higher rates may be achieved as long as the client feels that the product and service justify it;
- Competitiveness of an island destination with regards to prices mostly depends on the broader image the destination has worldwide. Whilst Maldives' tourism boom has its roots in diving and the likes, it is now considered a very up-market leisure destination and more tourists come to enjoy their honeymoon than come for diving. With markets permanently changing and evolving it is difficult to provide general answers. Value-for-money accommodation and its importance largely depends on supply and demand;
- For island based tourism the choice of comfortable transport is vital. Especially for small islands. Some tourists feel marooned and claustrophobic on small islands. The availability of transport eases these fears;
- Only some segments of the market are price sensitive. For the majority of travellers what is more important is value for money. Those with disposable income are willing to pay for 'experiences' more than the physical product.

5.4.5 Destination management factors

Analysis of survey data on *Price Factors* showed that the Maldives was performing well in the following important competitiveness factors (see quadrant one, Fig. 5.5):

35. DMO involvement in marketing
36. Resources for marketing
37. Environmental protection

The following factors were considered less important but well-performed in the Maldives (see quadrant two, Fig. 5.5):

38. Preparedness for crises
39. Cooperation in image-control

The following factors were considered less important and less-well-performed in the Maldives (see quadrant three, Fig. 5.5):

40. Public–private sector cooperation
41. Financial support for marketing
42. Local community involvement

The following factor was considered more important and less-well-performed in the Maldives (see quadrant four, Fig. 5.5):

43. Policies that support profitability
44. Hospitality training opportunities

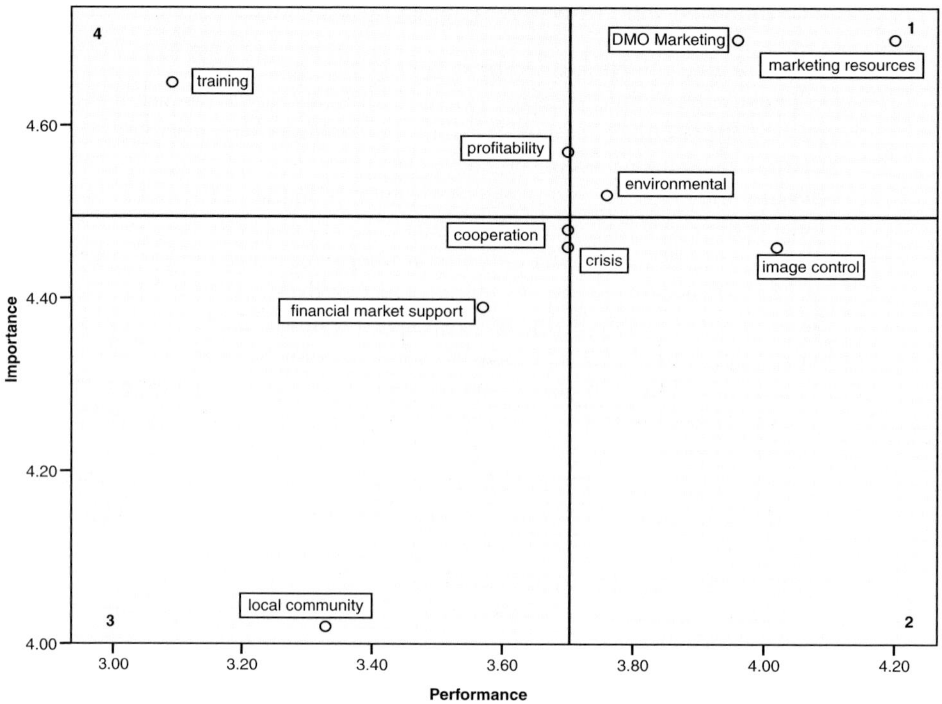

Fig 5.5. Destination management factors.

The following comments were returned by respondents:

- Ecology-friendly policies are important. Sustainable use of resources, respect for environment, pole and line fishing in Maldives, for example;
- Profit is not everything. Profit could be increased at the expense of a destination's short-term gain, e.g. Canary Islands, Seychelles. With regard to training, the continued use of the Hotel School in 1987 to 2007 is proof that training did not play a major role in increasing our competitiveness. I hereby acknowledge the initiatives by private sector to upgrade and develop their staff. I believe rather than providing more training courses, the industry should hire qualified expatriates to deliver a quality product and train staff on the job;
- The ideal working arrangement is PPP (public–private partnerships) that are meaningful. Focused national strategies that encourage ownership by the people are absolutely necessary for success;
- Through the many multipliers of tourism it affects the economy at many different levels. The flow through a destination's economy cannot be even or equitable unless the public and private sectors work together towards common national or regional goals.

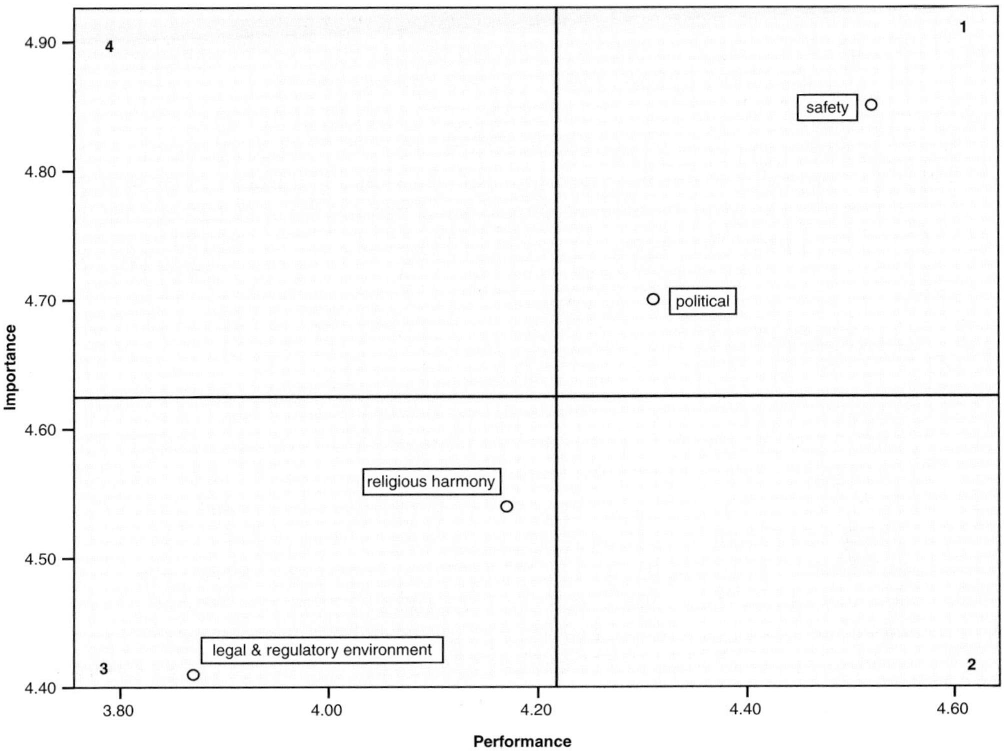

Fig. 5.6. Other factors.

5.4.6 Other factors

Analysis of survey data on *Related and Supporting Factors* showed that the Maldives was performing well in the following important competitiveness factors (see quadrant one, Fig. 5.6):

45. Safety and security
46. Political stability in destination

The following factors were considered less important and less well-performed in the Maldives (see quadrant three, Fig. 5.6):

47. Religious and social harmony
48. Business-friendly legal and regulatory environment

The following comments were returned by respondents:

- Because of the isolation Maldives provide to visitors (dedicated islands as resorts) a certain amount of instability [evidenced in the capital island of Malé] does not seem to affect [tourism on the other islands of] Maldives;

- Safety has become a key issue – clients have become more adventurous and perhaps a little more desensitized but the impact of a travel ban/advisory from a foreign office still has severe long-term consequences;
- Extremism and terrorism are the scourges of the 21st century. To be successful they must not be allowed to raise their heads in the destination;
- Above all, the flexibility to adopt and align products and to create new products with changes in demand will make the difference between success and failure in the foreseeable future.

5.5 Conclusions

Like many island tourism destinations, a combination of physical and environmental factors provide the basis for sustainable competitive advantage and these need to be continually well managed if the destination is to retain its competitive position. Factors, including the naturally endowed competitive advantage factors of climate and marine environment, provide the physical setting for water-based activities, supported by a friendly host community and services delivered by skilled and friendly staff at the resorts. The focus on marine activities has diluted the importance of other factors generally considered to be part of the tourism experience, such as culture, night-life, shopping and dining. Destination managers in the Maldives have recognized its unique geography, the smallness of the islands, the chains of atolls, the surrounding coral reefs with magnificent under-sea life as the competitive strengths of the destination which cannot be replicated elsewhere (as these are natural endowments), and have focused on promoting and delivering services around these features through cooperative marketing and management. However, as different market segments have different needs, DMOs must make constant evaluation of visitor-preferences and cater to these. For example, Russian and Chinese segments of visitors to the Maldives require different emphasis on, say, water-based activities, as compared to the importance placed on, say, the range of dining and night-time activities that may be required by other international visitors.

There are areas where the Maldives could improve their competitive performance, which centre around the currently limited facilities at the main international gateway airport catering to an ever-growing number of tourist arrivals. Overall, the destination remains competitive with respect to price of air travel and accommodation and is providing value for money to guests, however facilities and services at the airport itself remain under-improved. Destination managers need to maintain the high level of cooperation amongst industry participants, with respect to marketing and management, with particular attention provided to protection and upkeep of the natural environment, which in turn provide the basis for sustainable competitive advantage. Another area that needs attention is, most importantly, the provision of tourism and hospitality training in the Maldives in order to cater to changing needs that occur with changes in market segments. Linked to this is the necessity for the timely provision of government policies that support profitability, hence there is an opportunity for enhancing competitive advantage through encouraging higher levels of service quality linked with professional sales techniques that improve revenue and profitability.

The continued competitive performance of the Maldives can be attributed to the successive tourism master plans for the Maldives since 1983 that have embraced the principles of sustainable development rigorously throughout, and subsequent environmental planning guidelines that have stipulated the protection measures that developers must strictly follow. Further, the country's National Development Planning process formulated in cycles of 5 years emphasizes the importance of sustainable development and provides the aegis under which policies for development must proceed (Zulfa and Carlsen, 2011).

References and Further Reading

Buhalis, D. (2000) Marketing the competitive destination of the future. *Tourism Management* 21, 97–116.

Crompton, J. and Duray, N. (1985) An investigation of the relative efficacy of four alternative approaches to importance-performance analysis. *Journal of the Academy of Marketing Science* 13(4), 69–80.

Crouch, G.I. (2006) Destination competitiveness: insights into attribute importance. *International Conference of Trends, Impacts and Policies on Tourism Development*, Hellenic Open University, Heraklion, Crete.

Crouch, G.I. and Ritchie, J.R.B. (1999) Tourism, competitiveness, and societal prosperity. *Journal of Business Research* 44, 137–152.

Dwyer, L. and Kim, C. (2003) Destination competitiveness: determinants and indicators. *Current Issues in Tourism* 6(5), 369–414.

Enright, M.J. and Newton, J. (2004) Tourism destination competitiveness: a quantitative approach. *Tourism Management* 25, 777–788.

Enright, M.J. and Newton, J. (2005) Determinants of tourism destination competitiveness in Asia Pacific: comprehensiveness and universality. *Journal of Travel Research* 43(May), 339–350.

Enright, M.J., Scott, E.E. and Dodwell, D. (1997) *The Hong Kong Advantage*. Oxford University Press, Hong Kong.

Evans, M. and Chon, K. (1989) Formulating and evaluating tourism policy using importance-performance analysis. *Journal of Hospitality & Tourism Research* 13(3), 203.

Field, A. (2005) *Discovering Statistics using SPSS*, 2nd edn. SAGE Publications.

Gooroochurn, N. and Sugiyarto, G. (2005) Competitiveness indicators in the travel and tourism industry. *Tourism Economics* 11(1), 25–43.

Hassan, S. (2000) Determinants of market competitiveness in an environmentally sustainable tourism industry. *Journal of Travel Research* 38, 239–245.

Jick, T. (1979) Mixing qualitative and quantitative methods: triangulation in action. *Administrative Science Quarterly* 24.

Kline, P. (2000) *Handbook of Psychological Testing*. Routledge, London.

Liburd, J., Benkendorff, P. and Carlsen, J. (2011) Tourism and quality of life: How does tourism measure up? In: Sirgy, J., Purdue, R. and Uysal, M. (eds) *Handbook of Tourism and Quality-of-Life Research: Enhancing the Lives of Tourists and Residents of Host Communities*. Springer, New York, pp. 105–132.

Martilla, J.A. and James, J.C. (1977) Importance-performance analysis. *Journal of Marketing* 41, 77–79.

MOTAC (2011) *Maldives Tourism Yearbook 2011*. Ministry of Tourism, Arts & Culture, Maldives.

MOTAC (2013) *Fourth Tourism Masterplan* 2013–2017 Vol. 1, Strategic Action Plan. Ministry of Tourism, Arts & Culture, Maldives.

Oh, R. (2001) Revisiting importance-performance analysis. *Tourism Management* 22, 617–627.

Poon, A. (1993) *Tourism, Technology and Competitive Strategies*. CAB International, Wallingford, UK.

Porter, M. (1990) *The Competitive Advantage of Nations*. Free Press, New York.

Ritchie, J.R.B. and Crouch, G.I. (2000) The competitive destination: A sustainability perspective. *Tourism Management* 21(1), 1–7.

Ritchie, J.R.B. and Crouch, G. (2003) *The Competitive Destination. A Sustainable Tourism Perspective,* CABI, Wallingford, UK.

Swarbrooke, J. (2002) *The Development and Management of Visitor Attractions,* 2nd edn. Butterworth-Heinemann, Oxford, UK.

Zulfa, M. (2009) Stakeholder opinions of destination-competitiveness in small islands with particular reference to the Maldives. PhD Thesis, Curtin University, Perth, Western Australia.

Zulfa, M. and Carlsen, J. (2011) Planning for sustainable island tourism development in the Maldives. In: Carlsen, J. and Butler, R. (eds) *Island Tourism: Sustainable Perspectives.* CAB International, Wallingford, UK, pp. 215–227.

6 Quality of Life, the Tourism Area Life Cycle and Sustainability: A Case of Cuba

Meghan Beardsley*

Virginia Tech, USA

6.1 Introduction

Like all life on Earth, destinations go through a cycle experiencing different states of being over an expanse of years, sometimes plateauing for a prolonged period. Within the tourism discipline lens, Richard Butler (1980) shifted a concept into the forefront of the tourism literature that conceptually suggested the existence of a tourism area life cycle (TALC). Deriving some of its origins from ecological theory as well as the product life cycle, it was analogized that a tourist destination experiencing growth might develop without consideration of its future or preservation of its resources leading to issues of sustainability of the environment and its residents (Butler, 2006).

A potential and currently unexplored consequence of the progression of TALC and its correlation with a destination's sustainability is the resulting quality of life (QOL) of residents and tourists connected by the destination they share. Over the past decade, quality of life has become a research topic of significant interest for many in the field of tourism. Sometimes referred to as a state of well-being, the generic question related to QOL is: what is the impact of tourism on the destination's residents' and tourists' overall quality of life? Historically, QOL has been measured via a combination of objective and subjective measures with an understanding that there is a distinction between 'actual' QOL and an individual's perceived QOL, thus mandating the need for comparative information.

Destinations, regardless of their attractiveness, experience a life cycle, or, in other words, go through transitions. A transition could be political, economic, cultural, or environmental in origin and could be the catalyst that pushes a destination towards a different stage in a cycle. Cuba is one destination that is currently experiencing significant transitions in trade and international relations. A mere 93 miles from the

*E-mail: cinemeg2@vt.edu

US coast, Cuba functions as a Communist island nation seen by many as a 'must-visit' destination. Until recently, the possibility of US citizens visiting Cuba has been extremely unlikely. In mid-2015, Cuba made incredible strides forward in its trade relations with the USA, potentially opening doors to opportunities such as improving the overall quality of life of its citizens, but also will deal with sustaining itself. Being in such a transitional state, Cuba will be under the global microscope for years to come.

Now is an ideal time to observe Cuba and to assess where it is at in its TALC as well as taking a look at the current state of residents' QOL and comparing it at a future point once tourism has expanded forcing Cuba to develop more and more to accommodate demand. Being able to do so may allow for both or either the development of a correlation between QOL and TALC or even more robustly, the development of a life cycle attributed to the stages of resident and tourist QOL. Given TALC's connection with sustainability, it will be interesting to observe how this small island nation, which happens to be the largest in the Caribbean, will not only handle its political, trade and social transitions, but how it will plan and prepare for inbound tourists and how those plans impact the residents that news of the past couple of decades have reported to be unhappy and living Spartan-style existences.

This chapter intends to use the case of Cuba and its tourism status to provide preliminary support for the contention that the constructs of QOL, the TALC and sustainability are intertwined. After briefly exploring the background of each construct and any past linkages identified by researchers in the field, a case study of Cuba will be utilized to substantiate the notion that these constructs are interrelated. The case starts by first establishing the historical, political, economic and tourism setting of the nation. Following this, the positive and negative attributes of past efforts, present environment and future intentions for inbound Cuban tourism are explored. Applying Cuba's compounded background and tourism environment, warning signs are derived that could impact the three aforementioned constructs of this chapter in the future. The chapter concludes by offering propositions derived from the case study of Cuba and suggests potential future research endeavours for interested parties.

6.2 Foundation – Tourism Area Life Cycle

Butler brought the notion of the TALC to the forefront of the tourism literature in 1980, indicating influence from the resort cycle, tourism developing in three stages and the relationship between tourist types and destinations. The TALC model cites its origins from the product cycle concept where a product's sales ascend slowly, go through rapid growth, stabilize, then decline (Butler, 2006). The same process can be attributed to a tourist destination to explain the flow of visitors. The model suggests that each stage of the TALC can be facilitated by the level of marketing, information dissemination and other similar actions (Butler, 2006). The aim of the model is to identify at what stage a country is, based on a number of variables. From there, strategic decisions can be implemented based on their acquired knowledge of the model and past observations. Butler's TALC has been applied to over 30 country cases resulting in a wide range of success (Bolaky, 2008). De Albuquerque and McElroy successfully employed the TALC model to 23 Caribbean island nations in the 1990s and later to 32 member countries of the Caribbean Tourism Organization (Bolaky, 2008).

The TALC is divided into six stages (Butler, 2006). The first stage is the exploration stage consisting of small numbers of tourists, travelling independently and at irregular frequencies. The second stage is the involvement stage, where in response to an increase in visitors, local residents will develop facilities devoted to visitors and a market area is defined. Additionally, there is an increase in the level of organization involved in travel arrangements, which leads to governmental pressure to provide transportation and improve infrastructure. The third stage is the development stage (Butler, 2006; Bolaky, 2008). At this point, the tourist market area is well-defined, but a local's involvement and control over development is believed to decrease. Larger, updated tourism-devoted facilities will be raised. The fourth stage is the consolidation stage. Destinations will start to experience a decline in the rate of increase in visitors, and it is likely the number of visitors will outweigh the number of local residents (Butler, 2006; Bolaky, 2008). Also at this stage, the location's economy will be connected to tourism. Residents will start to resent the tourism-related facilities. Few additions are made during this stage. The fifth stage is the stagnation stage where the number of visitors will peak. Carrying capacity may be hit and surpassed at this point. The destination will rely on repeat visitation. The sixth, and final, stage is the decline stage. The area will reach a level of maturity that makes it less attractive and competitive compared to its adversaries. Local involvement might increase again for a number of reasons. Alternatively, the sixth stage could also be rejuvenation wherein if there is the addition of a man-made attraction or a previously unutilized resource is implemented, the area might experience renewed interest from visitors (Butler, 2006). The TALC continues to be an area of interest for academicians and practitioners alike.

6.3 Quality of Life

What is the impact of tourism on residents and tourists overall QOL? Quality of life, sometimes referred to as an individual's state of well-being, is a fairly new construct introduced into the tourism literature. Quality of life is often characterized by life domains that are impacted, which have yet to be clearly defined in a mutually agreed upon way. The most common domains employed in research are: economic (financial welfare), socio-cultural and physical/environmental. Importantly, researchers have recognized the necessity for triangulating their observation and measurement of the QOL construct. In an extensive review of QOL studies in the tourism literature, the compiled studies supported the notion that there is more than one way to measure QOL effectively (Uysal *et al.*, 2016). The measurement of QOL has been dichotomously broken down into 'actual' QOL and perceived QOL. The former looks at objective measures of the destination such as income, physical health, standard of living and crime rate (Uysal *et al.*, 2016). The latter looks at subjective indicators such as satisfaction with family, with work, with social life, with the community and with life overall (Uysal *et al.*, 2016).

6.4 Sustainability

Growing in prevalence in the 1990s, sustainability became an issue for discourse that started 'to direct the economic and political structures that constitute the present larger context of the tourism system, the industry, and its development' (Bramwell and Lane, 1993; Mowforth and Munt, 1998). Sharing similarities with carrying capacity, many critics feel that the research regarding sustainability is nothing new, just repackaged with a new name (Butler, 1999). Concerns for sustainable management of a destination arose when people started to realize the negative impacts tourism had on the environment and community. The precise, operational definition for sustainable tourism does not yet exist in the literature. There appear to be two popularly referenced identities for sustainable tourism. The first uses a broad approach to define the term as 'tourism which is economically viable but does not destroy the resources on which the future of tourism will depend, notably the physical environment and the social fabric of the host community' (Swarbrooke, 1999: 13). Those that subscribe to this broader perspective focus more on the sustainable use of resources (Hardy *et al.*, 2002). The second popular interpretation of sustainable tourism is to instead refer to it as the sustainable development in tourism and is made up of the ethical aspects of sustainability but does not use a tourism-centric approach when making development decisions (Butler, 1999; Burns, 1999). Hunter (1997) takes this forward by suggesting the interpretation of sustainable development is ascertained by how tourism plays a role in its development. The first role type is to carry out sustainable development through a 'tourism imperative' where the approach focuses efforts on the tourists and educating them (Hunter, 1997). The second role is through 'product-led tourism' promoting the need for creating new tourism products while simultaneously maintaining older tourism products (Hunter, 1997). Last, the third role is through 'environment-led tourism' wherein the approach suggests that strategic decisions are made in favour and protection of the environment (Hunter, 1997). Farrell (1992) presents a middle-of-the-road interpretation stating that sustainable development is the effort made to assume a balance in the development system between economy, environment and society, in such a way that none of those three is treated as more significant than the others.

Regardless, sustainable tourism development seems to be somewhat of a paradox, for it evidently endeavours to not overuse resources in order to continue to survive efficiently, but at the same time it is merely a means to use resources. For the sake of the chapter, given the varying definitions and opinions for sustainability and sustainable tourism, the interpretation is kept rather broad and more in line with Farrell's interpretation in order to allow for broad consideration when looking at the case for Cuba. There are many items to 'sustain' as it were. In the system of tourism in regards to destination decisions, there may be concerns to sustain profits, culture, lifestyles, the physical environment, and in some cases even a form of government, like with our future case of Cuba. Cuba is a socialist nation living in a predominantly capitalist world.

6.5 Relationship between Tourism Area Life Cycle and Quality of Life

Quality of life has grown in popularity within the tourism literature. Very few studies have endeavoured to connect the TALC with QOL. The relationship, however, seems fairly intuitive. When following along the TALC, the amount of visitors versus residents alternates as does the level of involvement and satisfaction with the destination in question. Therefore, the assumption that there is some level of correlation between the two is rather evident. In a study conducted in 1988, a notion was supported that tourism development can benefit a host community early on in development, but later decline as the development continued (Allen *et al.*, 1988). This study is one of the few to properly suggest the possibility of a connection between the TALC (i.e. earlier stages versus later stages) and the QOL of residents, although the latter term was not specifically used. In a review of QOL, this linkage was further supported with collective findings that identified that the level of tourism development affects the residents' attitude toward economic, socio-cultural and environmental factors (Uysal *et al.*, 2016). Three conclusions can be made from this review of a collection of studies. The first is that the impact of tourism development does play a significant role for the host community residents' QOL (Uysal *et al.*, 2016). The second is that not all residents perceived tourism's impact similarly, which could suggest that their type of stakeholder role could be moderating this relationship (Uysal *et al.*, 2016). Third, like the 1988 study mentioned, the authors similarly concluded that the residents' perceptions and attitude toward tourism (and QOL level) tend to differ depending on the level of tourism development with most of the benefits accruing in the earlier than later stages of development (Uysal *et al.*, 2016). There are potentially other studies out there that have inadvertently supported this relationship as well.

6.6 Relationship between Tourism Area Life Cycle and Sustainability

Very little effort in terms of research has been made to understand the relationship between the TALC and sustainability. One study that started to touch on that notion proposed a multiplicative approach through a case study. Within this study, the authors suggested that the TALC model be used collaboratively with the teleological model to explain any movement towards sustainability of an island destination in its maturity stage (Oreja-Rodríguez *et al.*, 2008). The teleological model suggests that the evolution of a destination is conditioned by a range of variables including institutional decision making, identifying objectives, strategic planning and social construction (Oreja-Rodríguez *et al.*, 2008). This study alludes to connection with sustainable development. More endeavours are needed for understanding the relationship involved between sustainability and the TALC, for it might help a destination understand where they should be at with their sustainable development efforts.

6.7 Relationship Between Quality of Life and Sustainability

Like the relationship between TALC and sustainability, few studies have explored the relationship between QOL and sustainability. Most analysis supporting the relationship has emerged within the past decade. A couple of studies presented empirical support for the relationship between QOL and sustainability. The first found that perceived economic benefits, environmental sustainability and perceived social costs positively affect residents' QOL (Jeon *et al.*, 2016). A second study found that the cultural and environmental effects of tourism affected residents' QOL and also demonstrated characteristics that affect their perceptions of tourism domains (Khizindar, 2012). Choi and Sirakaya (2005) were the first to attempt to develop a scale that measured residents' attitudes towards sustainable tourism. This scale is colloquially referred to as SUS-TAS (Choi and Sirakaya, 2005). These authors have continued to build upon the validity and reliability of this scale with other studies that updated on the original scale (Sirakaya-Turk *et al.*, 2008; Hung *et al.*, 2011). One later study was able to use SUS-TAS to find that environmental sustainability and perceived economic benefits have a direct effect on residents' QOL (Chancellor *et al.*, 2011). In a well-rounded review of the quality of life literature, Uysal *et al.* (2016) were the first to suggest an interrelationship between sustainability, tourism, and quality of life.

6.8 Background and Context

Cuba is an island nation located between the Caribbean Sea and the North Atlantic Ocean approximately 93 miles south of Key West, Florida and 48 miles from Haiti (Nikolova, 2011; CIA, 2015). Slightly smaller than the state of Pennsylvania, Cuba's land area is 109,820 km^2 and is the largest country in the Caribbean and the westernmost island of the Greater Antilles (CIA, 2015). With an estimated population of 11,259,000 as of 2015, the country's capital city of Havana harbours 2,146,000 residents, which is about 19% of the total population (UN, 2015). Of this reported population, 64.1% are Caucasian, 26.6% are mestizo, and 9.3% are black (CIA, 2015). In order to be characterized as earning an upper middle income, a Cuban must earn anywhere between US$2,946 and US$12,195 per year according to the gross national income per capita (World Bank, 2009).

Over the past 60 years, Cuba has experienced several political upheavals eventually leading to Fidel Castro coming to power allowing for the growth of Communism. As a result of Cuba's communist practices and friendly relationship with the Soviet Union during the 1960s, the USA executed a commercial, economic and financial embargo on Cuba. Simultaneously, Cuba was suspended from the Organization of American States. Following the collapse of the Berlin Wall and the loss of former Soviet subsidies worth US$4–6 billion annually, Cuba experienced a severe economic downturn in 1990. The World Bank reported that Cuba's GDP was $77.15 billion in 2013. The three contributory categories of Cuban GDP were: agriculture (4%), manufacturing/industrial (22.4%) and services (73.5%). Cuba's primary industries are: sugar, manufacturing, other agriculture and services. In 2010, it was reported that most of the industries in the nation are owned and controlled by the state, which is responsible for employing approximately 78% of the country's people (Economy Watch, 2010).

Common adjectives used to describe Cuba at present include words such as spartan, neglected and decaying. The socialism fist slammed down on its residents, crushing the potentiality and competiveness of the entrepreneurial spirit. This lack of freedom for society has led to the steady flow of refugees abandoning Cuban soil in search of asylum in countries such as the USA, Italy and Spain (Cook, 2014). As of 2013, 1.48 million people born in Cuba are now living in other countries. In the same year, 1.2 million Cuban migrants were reported to be living in the USA

(Pew Research Center, 2014). When considering these statistics it is important to remember that there is a percentage of Cuban refugees that may have entered the USA via raft and thus are unreported.

6.9 Cuba Transitions

Within the past decade, a 'thaw' has started to occur between the USA and Cuba. In 2011, the government held the first Cuban Communist Party Congress in 14 years in order to discuss economic change. This change would shift them closer towards a more market-oriented system. Additionally, the Cuban government has started to slowly implement some limited economic reform, now allowing its citizens to purchases electronics, stay in hotels, and register as self-employed/entrepreneurs (CIA, 2015). These changes now allow for private ownership and sale of real estate and products. For example, private farmers can now sell their wares directly to a hotel as a supplier. This alters who the relevant stakeholders are in Cuba, especially when it comes to the tourism industry.

In regards to political and international relations' transitions, Cuba and the USA held ceremonies in 2015 announcing intentions to re-establish diplomatic relations with the hopes that in the future, activities such as tourism will be commonplace. At present, US citizens can only visit if they fall under 12 highly regulated categories: family visits, official business of a government, journalistic activity, professional research/meetings, religious activities, educational activities, performances/competitions, humanitarian projects, activities of a private foundation, exportation/importation of information, certain authorized export transactions of products, and support for the Cuban people (Mathis-Lilley, 2015). Cruise lines have announced that they too will soon be allowed to drop anchor in Cuba and have started chartering trips for May 2016 (Burnett, 2015).

6.10 Cuba Tourism: The Good, the Bad and the Ugly

While in the midst of the 1990s economic crisis, the Cuban government elected to invest in tourism. As a result, they invested US$3.5 billion in their tourism industry, increasing the number of rooms available from 12,000 to 35,000 and within the decade the tourism sector multiplied its gross income eight-fold (Castillo and Gaspar, 2002; Nikolova, 2011). Infrastructural developments were also carried out, such as devoting resources to airports and causeways. Despite these efforts, tourism numbers did not increase as much as originally aspired.

Given Cuba's insular nature, this country's attempts for tourism and agricultural growth combat against its limited available resources and sustainable efforts. Cuba's massive deforestation and previous excessive freshwater withdrawals have threatened their sustainability in the past (Office of Global Analysis, FAS, USDA, 2008). The use of freshwater only decreased in the 1990s because of the ending of the relationship between Cuba and the Soviet Union that reduced their capacity for agricultural production (Office of Global Analysis, FAS, USDA, 2008). Still, like all island destinations, there will be a challenging battle for balance of resources as Cuba's tourism industry continues to grow with the influx of US visitors, which will counteract their efforts for sustainability.

Interestingly, according to Avella and Mills (1996, p. 55), 'for the Cuban people tourism is synonymous with economic, social, and environmental decline'. In a book that explored the resources, planning and environmental management of the Caribbean, the authors noted that there were numerous, significant constraints on Cuba's tourism development. The first is the amount of leakage. In 1997, the leakage for Cuba was reported between 70 and 75% suggesting that the economy was not successfully producing goods of a quality that was considered acceptable by international tourists (Haines, 1997; Avella and Mills, 1996). The second constraint identified was the poor infrastructure and quality of accommodation. During the 1990s, it was reported that the country's infrastructure was in such a state that big hotel operators would not invest in properties there. Additionally, the infrastructure was in need of major modernization, in particular transportation, roads, energy, water, sewerage and telecommunications (Barker and McGregor, 2003).

At the time, the goal was to achieve 2.5 million tourists by 2000, but Cuba did not succeed in getting this number until 2010 (Barker and McGregor, 2003; Nikolova, 2011; Table 6.1). The main visitor markets are Canada, the UK, Spain, Italy, Germany, France and Mexico (Nikolova, 2011). In addition to their infrastructural investments, Cuba endeavoured to adopt environmentally sustainable strategies and attributes those strategies as a leading influence on the increase in tourist numbers (Nikolova, 2011). As of 2010, Cuba reported being host to 2.5 million visitors. In 2014, the nation hit a new tourism figure rung, receiving approximately 3 million visitors (Dube, 2015).

By the end of the 1990s, Cuba had the second largest hotel capacity of the Caribbean accounting for 73% of the nation's investments in tourism (Castillo and Gaspar, 2002). The remaining 27% was intended for infrastructural development for especially airports, causing an imbalance of investments for other tourist-related items such as restoration and recreation (Castillo and Gaspar, 2002). This imbalance led to a persistent increase in hotel capacity investment and consequentially increased the lack of diversity in tourist attractions. This will eventually challenge the Cuban tourism stakeholders to reconsider how they form the tourism product (Castillo and Gaspar, 2002). By 2015, Cuba's ministry of tourism reported about 61,000 hotel rooms not including those offered by Airbnb (Burnett, 2015). As a result, there have numerous comments by hotel operators and tour agencies regarding Cuba's inability to accommodate the tourist demand, and it is the consensus that the issue is only going to become more severe (Burnett, 2015; Dube, 2015; Whitefield, 2015).

Following the January announcement of the 'thaw', the US Secretary of State John Kerry announced that US travel to Cuba was up 35% by August (Dube, 2015). There are still US enforced policies that restrict US corporations from entering the Cuban market, but there are hopes of this lessening in the future. According to Marriott International the island could receive 1.5 million visitors a year if US travel restrictions are removed, but its chief development officer for the Caribbean and Latin American region says 'Cuba's tourism infrastructure, in our opinion, is not ready to absorb that' (Dube, 2015). As one of the leading multinational hospitality corporations, this declaration should serve as a warning to the Cuban government and its residents. Cuba will find it challenging to keep up with its environmental sustainability standards when trying to accommodate the significant influx of air travel and hotel growth.

Table 6.1. Cuba's tourism industry, 2004–10 (in millions; Office of National Statistics, 2014).

	2004	2005	2006	2007	2008	2009	2010
International arrivals							
Visitors	2049	2319	2221	2152	2348	2430	2532
Tourists	2017	2261	2150	2119	2316	2405	2507
Excursionists	32	58	71	33	32	25	25
Cruise passengers	5	17	30	7	5	4	2
Visitor arrivals by region							
Africa	6	6	7	7	8	9	9
Americas	1026	1216	1149	1173	1380	1536	1664
Europe	977	1048	1014	924	909	838	810
East Asia and Pacific	34	42	44	42	44	41	41
Central Asia	4	5	5	4	5	4	6
Middle East	2	2	2	2	2	2	2
Departures	124	162	199	194	202	206	251

In a study commissioned by the United Nations, TALC was implemented to assess what stages Caribbean countries were at based on a variety of factors over 15 years prior to 2007. One of the variables included broad classes of tourism strategies. Cuba's strategy is believed to 'continuously rely on mass-tourism to build on comparative advantages of "sun, sand and sea", the presence of scale economies, all-inclusive packages, and large amounts of investment so as to build or maintain market share and revenues and move along in Stage 2 <investment> or Stage 3 <development>'. As of 2007, Cuba was between Stage 3 <development> and Stage 4 <consolidation>. Given their intentions to upgrade and update infrastructurally and superstructurally, Cuba has two potential routes it might experience. If they manage to catch up with tourist demand following the lightening of US travel regulations, Cuba may hit stagnation and will face the challenge of either being able to rejuvenate their destination or face decline. However, this seems like a threat for after 2025. Overall, it appears that adopting more market-oriented methods could help Cuba economically and its citizens as well.

6.11 Warning Signs for Cuba

Based on a review of statistics and news reports, Cuba is at risk in several ways regarding its sustainability, tourism industry and residents' QOL. Cuba is about to potentially experience such a significant increase in visitors that has never occurred in its history if the US travel regulations are lightened. The small island has prided itself in its authenticity and lack of outside influence, but with such growth and development this could be lost in translation. This has the potential to negatively impact both 'actual' QOL and perceived QOL if cultural pride, space, lack of pollution, low traffic and lack of overcrowding are highly valued by a local.

Another threat that is developing as a result of the increase in visitation and interest to visit, is to Cuba's carrying capacity. Cuba is struggling to keep up supply with the demand coming from the world. Due to the lack of supply, prices in rooms have increased and there is low availability during the high season making it difficult for people to visit. As Cuba endeavours to construct more hotels and thus increase the number of rooms available, more labour is required, thus increasing job opportunities for locals. The downside of these efforts though is that all this construction can harm the environment and impact Cuba's environmental sustainability efforts. All of this can lead to the negative impact on residents' and tourists' QOL in terms of the economy and environment.

Presently, hotels pay a dollar salary intended for their employees to a government department, who in turn converts this to a Peso wage generally around US$20 per month. This has led to a low labour quality and a higher threat of theft. Additionally, the struggle for paying jobs is so aggressive that the gap in the standard of living has significantly widened leading to a very active black market. An economic struggle between resentful locals and their employers and customers might occur and result in theft, corruption, violence and low service quality, all of which negatively impact tourism overall.

Cuba currently resides between stages two and three of the TALC, but based on this case the QOL of locals and tourists appears to be decreasing. Verily, Cuba's economy will experience an influx of money with the incoming cruise tourists and potential American visitors, but what is the cost to everyone? In line with the proposition that QOL experiences stages simultaneously with the TALC stages, local involvement has a curvilinear association with the cycle, and thus so could quality of life.

Several studies are needed as a result of this case. For example, a study must be carried out once the USA finally lifts the travel regulations, thus opening the gates for the 1.5 million people predicted to travel to Cuba. Another longitudinal study is needed to investigate Cuba's environmental sustainability management in the next 10 years given the anticipated influx of planes, cruise lines, hotel construction and foot traffic. Given the transitions taking place in Cuba, the relevant stakeholders are changing, and with that comes the call to investigate who these new stakeholders are, what their influence is, and what is important to them. Most importantly, efforts must be made to calculate the carrying capacity for Cuba and its popular attractions in order to properly manage supply and demand in a way that is sustainable. Lastly, once the QOL cycle is officially developed it should be used in consideration of Cuba once more to see where Cuba rests in the cycle.

6.12 Conclusions

Based on the brief conceptual background and the case of Cuba presented earlier, there are implications that a QOL cycle could be developed and grounded in the discipline of tourism. Naturally, more in-depth development and analysis is required to properly develop such a model. Proposing such a model seems obvious, but like Lundgren (1984: 22) said about TALC in response to its being intuitively obvious, 'Butler put into the realistic cyclical context a reality that everyone knew about, and clearly recognized, but had never formulated into an overall theory'. The same could be possible for the proposed QOL cycle if formulated appropriately.

In addition to this proposition, the other purpose of this chapter was to look at Cuba as a destination in transition and observe how that is impacting the QOL of citizens and tourists alike. There are several research considerations that come to mind in regards to this case study. First, when an island that is also viewed as an attractive tourism destination is going through a significant political, economic, or cultural transition, what decision-making takes place for handling infrastructural and superstructural developments? Given that tourism has the potentiality of being a major contributor to that island's economy, the stakeholders cannot afford to put tourism development on the backburner during a time of transition. In fact, this transition could determine where they fall on the TALC and forecast their fate overall once they hit a state of maturity. Furthermore, how do these aforementioned developmental endeavours contribute to/maintain/or negatively impact the quality of life for residents and tourists? Infrastructural developments are likely to be more beneficial for residents than the superstructural development. Also, these developmental efforts plus any results of the transitions could impact residential life and the experience of tourists. A longitudinal study that first interviewed residents and tourists prior to the transitions and continued forward until the present time could go a long way in understanding an island in transition, its developmental endeavours and the QOL for residents and tourists under such circumstances. Given the case study presented above, can the content serve as a means to make predictions regarding other nations that may experience transitions in the future? Likewise, the proposed QOL cycle could play a predictive role in this as well.

In tangent of the aforementioned research considerations, given that sustainability is a growing concern for countless destinations, a study that looks into the relationship between the TALC and sustainability could prove beneficial. It could help destinations to identify where they are in the cycle and see if their sustainable efforts are on par with what they should be in order to remain a sustainable destination. Finally, as it has been lightly touched on throughout this chapter, serious research efforts should be made into the development and analysis of the concept of the QOL cycle and its connections to tourism and sustainability.

This case study was constructed with the intent to support the intuitive notion of an interrelationship between the TALC, QOL and sustainability. Furthermore, it provided preliminary, evidentiary support at the possible existence of a QOL cycle that is potentially correlated with the TALC. The next leap from this research is to build upon this preliminary support and endeavour to develop and potentially operationalize the proposed scale. Lastly, given the unique characteristics of this case study, perhaps it will prove useful for other popular destinations that are experiencing significant transitions for understanding the potential threats and opportunities they might experience and how that could impact both their residents and tourists.

References

Allen, L.R., Long, P.T., Perdue, R.R. and Kieselbach, S. (1988) The impact of tourism development on residents' perceptions of community life. *Journal of Travel Research* 27(1), 16–21.

Avella, A.E. and Mills, A.S. (1996) Tourism in Cuba in the 1990s: back to the future? *Tourism Management* 17(1), 55–60.

Barker, D. and McGregor, D.F. (2003) *Resources, Planning and Environmental Management in a Changing Caribbean.* University of West Indies Press, Jamaica, West Indies.

Bolaky, B. (2008) Tourism life cycle, tourism competitiveness and upgrading strategies in the Caribbean. ECLAC subregional headquarters for the Caribbean (studies and research papers), Port-of-Spain, Trinidad and Tobago.

Bramwell, B. and Lane, B. (1993) Sustaining tourism: An evolving global approach. *Journal of Sustainable Tourism* 12, 495–529.

Burnett, V. (2015) The latest on travel to Cuba. *The New York Times.* Available at: http://www.nytimes.com/2015/01/21/travel/the-latest-on-travel-to-cuba.html?_r=0 (accessed 1 December 2015).

Burns, P. (1999) Paradoxes in planning: Tourism elitism or brutalism? *Annals of Tourism Research* 26, 329–348.

Butler, R. (1980) The concepts of a tourist area life cycle of evolution: Implications for management of resources. *Canadian Geographer* 24(1), 5–12.

Butler, R. (1999) Sustainable tourism: A state-of-the-art review. *Tourism Geographies* 1, 7–25.

Butler, R. (2006) *The Tourism Area Life Cycle,* Vol. 1. Channel View Publications, Clevedon, UK.

Castillo, O.G. and Gaspar, N.G. (2002) *Tourism.* Available at: http://revista.drclas.harvard.edu/book/focus-cuba-and-caribbean (accessed 1 December 2015).

Chancellor, C., Yu, C.S. and Cole, S.T. (2011) Exploring quality of life perceptions in rural midwestern (USA) communities: An application of the core–periphery concept in a tourism development context. *International Journal of Tourism Research* 13(5), 496–507.

Choi, H.S. and Sirakaya, E. (2005) Measuring residents' attitude toward sustainable tourism: Development of sustainable tourism attitude scale. *Journal of Travel Research* 43(4), 380–394.

CIA (2015) *The World Factbook: Cuba.* Available at: https://www.cia.gov/library/publications/the-world-factbook/geos/cu.html (accessed 1 December 2015).

Cook, L. (2014) Here's where the Cuban refugees live. *US News.* Available at: http://www.usnews.com/news/blogs/data-mine/2014/12/17/heres-where-the-cuban-refugees-live (accessed 1 December 2015).

Dube, R. (2015) Cuban tourism is a work in progress. *The Wall Street Journal,* 16 September. Available at: http://www.wsj.com/articles/cuban-tourism-is-a-work-in-progress-1442437838 (accessed 1 December 2015).

Economy Watch (2010) Cuba Industry Sectors. Available at: http://www.economywatch.com/world_economy/cuba/industry-sector-industries.html (accessed 1 December 2015).

Farrell, B. (1992) Tourism as an element in sustainable development: Hana, Maui. In: Smith, V. and Eadington, W. (eds) *Tourism Alternatives.* University of Pennsylvania Press, Philadelphia, Pennsylvania, pp. 115–134.

Haines, L. (1997) Reassessing Cuba: emerging opportunities and operating challenges. *EIU Special Report.* Economist Intelligence Unit, London.

Hardy, A., Beeton, R. and Pearson, L. (2002) Sustainable tourism: An overview of the concept and its position in relation to conceptualisations of tourism. *Journal of Sustainable Tourism* 10, 475–496.

Hung, K., Sirakaya-Turk, E. and Ingram, L.J. (2011) Testing the efficacy of an integrative model for community participation. *Journal of Travel Research* 50(3), 276–288.

Hunter, C. (1997) Sustainable tourism as an adaptive paradigm. *Annals of Tourism Research* 24(4), 850–867.

Jeon, M.M., Kang, M.M. and Desmarais, E. (2016) Residents' perceived quality of life in a cultural-heritage tourism destination. *Applied Research in Quality of Life* 11(1), 105–123.

Khizindar, T.M. (2012) Effects of tourism on residents' quality of life in Saudi Arabia: an empirical study. *Journal of Hospitality Marketing & Management* 21(6), 617–637.

Lundgren, J.O. (1984) Geographic concepts and the development of tourism research in Canada. *Geojournal* 9, 17–25.

Mathis-Lilley, B. (2015) Prior approval no longer required as of Friday for 12 types of travel to Cuba. *Slate.* Available at: http://www.slate.com/blogs/the_slatest/2015/01/15/cuba_travel_categories_12_types_of_visit_legally_allowed_without_prior_permission.html (accessed 1 December 2015).

Mowforth, M. and Munt, I. (1998) *Tourism and Sustainability: A new tourism in the third world.* Routledge, London.

Nikolova, M.S. (2011) Travel and tourism – Cuba – November 2011. *Mintel Reports.* Available at: http://academic.mintel.com.ezproxy.lib.vt.edu/display/601655 (accessed 1 December 2015).

Office of Global Analysis, FAS, USDA (2008) Cuba's Food & Agriculture Situation Report (pp. 1–64). Available at: https://www.ilfb.org/media/546435/fas_report_on_cuba.pdf (accessed 1 December 2015).

Office of National Statistics (2014) *Statistical Yearbook of Cuba: Tourism.* Available at: http://www.one.cu/aec2014/15%20Turismo.pdf (accessed 1 December 2015).

Oreja-Rodríguez, J., Parra-López, E. and Yanes-Estévez, V. (2008) The sustainability of island destinations: Tourism area life cycle and teleological perspectives. The case of Tenerife. *Tourism Management* 29(1), 53–65.

Pew Research Center (2014) Origins and Destinations of the World's Migrants. Available at: http://www.pewglobal.org/2016/05/17/global-migrant-stocks (accessed 1 December 2015).

Sirakaya-Turk, E., Ekinci, Y. and Kaya, A.G. (2008) An examination of the validity of SUS TAS in cross-cultures. *Journal of Travel Research* 46(4), 414–421.

Swarbrooke, J. (1999) *Sustainable Tourism Management.* CAB International, Wallingford, UK.

UN (2015) United Nations Statistics Division. Country profile: Cuba. Available at: http://data.un.org/CountryProfile.aspx?crName=cuba#Economic (accessed 1 December 2015).

Uysal, M., Sirgy, M.J., Woo, E. and Kim, H.L. (2016) Quality of life (QOL) and well-being research in tourism. *Tourism Management* 53, 244–261.

Whitefield, M. (2015) Full house: Cuba tour operators struggle to find enough hotel rooms. *Miami Herald*, 12 November. Available at: http://www.miamiherald.com/news/nation-world/world/americas/cuba/article44510043.html (accessed 1 December 2015).

World Bank (2009) Cuba: World trade indicators. Available at: http://documents.worldbank.org/curated/en/2009/12/17002907/cuba-world-trade-indicators-2009-trade-glance (accessed 1 December 2015).

7 Place Satisfaction, Place Attachment and Quality of Life: Development of a Conceptual Framework for Island Destinations

HAYWANTEE RAMKISSOON*

School of Marketing, Curtin University and Behaviour Works Australia, Monash University, Australia

7.1 Introduction

The success of tourism in island destinations depends largely on their distinctive cultural and natural assets, which set the context for the psychological exchanges between the visitors and the physical space (Ramkissoon, 2015). The socially constructed meanings visitors ascribe to this space transforms it into a place (McKercher *et al.*, 2015; Ramkissoon and Mavondo, 2015) when visitors endow it with value (Tuan, 1980; Ramkissoon *et al.*, 2013a).

Small island economies are associated with their small size, insularity, remoteness and proneness to natural disasters (Ramkissoon and Uysal, 2011; Cheng *et al.*, 2013) and as such considered by many in a disadvantageous position in global competition (Venegas Sr and Croes, 2003). However, tourism in small islands played a significant role in overcoming these constraints, and is now a common strategy to achieve greater economic performance (Croes, 2006; Hoti *et al.*, 2007). Competition between places forced island destinations to promote their distinctiveness to attract and retain local and international visitation. In addition to the traditional sand, sun and sea (3S), islands host unique natural and cultural sites such as botanical gardens, marine parks, islets, natural landscape and cultural sites (Ramkissoon and Uysal, 2010; Ramkissoon *et al.*, 2011). These value-added island resources contribute to the core 3S product (Ramkissoon, 2015) and may increase place satisfaction for visitors in search of distinctive places. With increasing visitation at the ecological and cultural

*E-mail: haywantee.ramkissoon@curtin.edu.au

environments (Gossling, 2003; Lee *et al.*, 2010), it is essential to manage these special island resources for sustainable tourism. It is also vital from a marketing perspective to maintain destination competitiveness (Buhalis, 2000; Rodriguez *et al.*, 2008).

Literature evidences that satisfaction with natural and cultural places could enhance visitors' levels of place attachment (Ramkissoon and Mavondo, 2014; Ramkissoon *et al.*, 2014). Place attachment is conceptualized as the extent to which an individual values or identifies himself/herself with a specific environment (Ramkissoon *et al.*, 2013a). In other words, place attachment depicts the emotional ties a person shares with a place. Originating from the attachment theory (Bowlby, 1980a) depicting the mother–infant bond, place attachment is an important indication of place affect, place identity, place dependence and place social bonding (Ramkissoon *et al.*, 2012, 2013a, b) in natural and cultural settings.

Some evidence further suggests that place attachment to natural and cultural settings may lead visitors to engage in pro-environmental behaviours. Halpenny (2010) and Kelly and Hosking (2008) reported place attachment to be significantly associated with pro-environmental behaviours in Canada and Australia. Similar findings were reported by Cheng *et al.* (2013) in Penghu island in Taiwan. Further, associations between place attachment, with dimensions of place dependence, affect, social bonding and visitors' low and high pro-environmental behaviours were noted in natural settings (Ramkissoon *et al.*, 2013a).

While some studies have investigated the associations between place satisfaction, place attachment and pro-environmental behaviour in tourism (e.g. Uzzell *et al.*, 2002; Ramkissoon *et al.*, 2014; Ramkissoon and Mavondo, 2015), findings are far from being conclusive. Researchers are delving further for clarifications on the association of these constructs. This study provides further explanation about how place satisfaction and pro-environmental behaviour are linked. It proposes that place attachment (with dimensions of place affect, identity, dependence and social bonding) mediates the relationship between place satisfaction and pro-environmental behaviour at natural and cultural attractions in island destinations.

Another important gap that remains to be addressed is how the above-mentioned constructs affect visitors' overall quality of life, which in turn may have an impact on future behaviour (Mannell and Kleiber, 1997; Kim *et al.*, 2015). Place attachment has been proposed as an antecedent to behavioural intentions (e.g. Veasna *et al.*, 2013; Ramkissoon, 2015). This study proposes that pro-environmental behaviour is one of the key mediators in the relationship between place attachment and quality of life. An understanding of the link between visitors' place attachment and quality of life would allow island destination managers and marketers to improve their efforts in developing, marketing and managing natural and cultural resources. This will yield sustainable benefits to the tourism industry and the community.

Although several studies have investigated the association between visitors' travel experiences on their quality of life, research on the links between place satisfaction, place attachment (with dimensions of affect, identity, dependence, social bonding), pro-environmental behaviour and quality of life is scant. Challenging traditional discourses on behavioural modelling, this study offers to literature a new model with integrative concepts in small island economies. It contributes to theoretical advancement of knowledge. Important practical implications for sustainable tourism in island destinations are discussed.

7.2 Literature Review

7.2.1 Place satisfaction

Satisfaction refers to the consumer's response to attitudes including judgements following a purchase or a series of consumer–product interactions (Lovelock and Wirthz, 2007). It is considered a critical element in measuring customer experience with a product (Tudoran *et al.*, 2012; Rahul and Majhi, 2014), which park managers are increasingly seeking to achieve (Crilley *et al.*, 2012; Ramkissoon *et al.*, 2014). While a plethora of studies on customer satisfaction has focused exclusively on travel needs (e.g. Uysal and Williams, 2004; Uysal, 2006), more research is needed exploring visitors' place satisfaction with natural and cultural attractions in island destinations. This has implications for sustainable island tourism from both a management and marketing perspective.

Place satisfaction and place attachment in island states

Place satisfaction with natural and cultural attributes at a destination has been defined as the perceived quality of a setting, meeting the visitor's needs for the physical characteristics and services (Ramkissoon *et al.*, 2012). As with other destinations, visitor place satisfaction and place attachment in island economies are vital to the success of their tourism industry relying on unique nature-based and cultural tourist attractions. A number of studies found significant relationships between place satisfaction and place attachment at tourist destinations. While several studies have attempted to identify place attachment as a predictor of visitor satisfaction (e.g. Yuksel *et al.*, 2010; Ramkissoon *et al.*, 2013a, b; Veasna *et al.*, 2013), a recent study by Ramkissoon and Mavondo (2015) suggests visitor satisfaction as an antecedent of place attachment. Lee *et al.* (2012) found a direct positive effect of festival satisfaction on festival attendees' place dependence and identity/social bonding. Halpenny (2006) concluded that an individual's level of satisfaction with Point Pelee's national park in Canada had a positive effect on overall place attachment.

Visitor satisfaction and pro-environmental behaviour

Pro-environmental behaviour is defined as the willingness to undertake actions to minimize environmental harm and improve environmental conditions of a setting (Ramkissoon *et al.*, 2012). Research suggests that the relationship between place satisfaction and visitors' pro-environmental behaviour in natural settings is dynamic and complex (Ramkissoon *et al.*, 2013b). Stedman (2002) argued that respondents with lower levels of satisfaction showed greater willingness to act to counter environmental changes, particularly those who were displeased with physical conditions such as the water quality, scenery and atmosphere. Ramkissoon *et al.* (2013b) found a significant positive effect of visitors' place satisfaction on their low effort pro-environmental behavioural intentions but no significant relationship with their high pro-environmental behaviour. Halpenny (2006) found no significant relationship between park visitors' satisfaction and their pro-environmental behavioural intentions in a Canadian national park context. Several other studies have demonstrated positive correlations between satisfaction and pro-environmental behaviours (e.g. Baral *et al.*, 2008; Davis *et al.*, 2011; Lopez-Mosquera and Sanchez, 2011; Paille and Mejia-Morelos, 2014). However, findings remain inconclusive, which may suggest that a mechanism linking visitors' place satisfaction with their pro-environmental behaviour is missing. There may be mediators that were omitted in prior studies. It is suggested that one of these missing links is place attachment. With visitors' quest for distinctive cultural and natural resources (Ramkissoon and Uysal, 2011; Ramkissoon, 2015), researchers and destination marketers are progressively more interested in furthering the understanding between visitors' place satisfaction and their pro-environmental behaviours in island destinations.

7.2.2 Place attachment as a mediator

In the environmental psychology literature, place attachment refers to the bonding people share with places (Raymond *et al.*, 2011; Ruiz and Hernandez, 2014). Place attachment emerges as people view a place with confidence where they can engage in activities (Hidalgo and Hernandez, 2001). Amongst the plethora of terms describing the relationship between people and spatial settings, place attachment is the most popular (Ramkissoon *et al.*, 2012). It has received considerable attention as an integrative construct (e.g. Yuksel *et al.*, 2010; Veasna *et al.*, 2013). While the most commonly studied dimensions are place dependence (Stokols and Shumacker, 1981) and place identity (Vaske and Kobrin, 2001; Walker and Chapman, 2003), place affect (Halpenny, 2010; Ramkissoon *et al.*, 2012, 2013c) and place social bonding (Wynveen *et al.*, 2012) are receiving increasing attention.

A body of research in nature-based settings has recently recognized the associations between place satisfaction, place attachment and pro-environmental behaviour. Findings on the association between these constructs remain inconclusive (see Halpenny, 2010; Ramkissoon *et al.*, 2013b) suggesting that place attachment (dependence, identity, affect, social bonding) might be the missing mechanism linking visitor satisfaction to pro-environmental behaviour. It is argued that the impact of visitors' satisfaction levels on their pro-environmental behaviours would be enhanced by their levels of place attachment. Studies investigating the mediating effects of place attachment with the four dimensions of place affect, identity, dependence and social bonding at cultural and natural island attractions, are yet to be explored in the tourism literature.

Place affect

To further enhance the conceptualization and measurement of place attachment, researchers have included the place affect dimension (Rolero and De Picolli, 2010; Ramkissoon *et al.*, 2012) to existing well-established dimensions of place dependence and place identity. Place affect is defined as the emotional bond individuals develop with environmental settings (Halpenny, 2010) by building their sentiments about the place (Tuan, 1977). Hinds and Sparks (2008) concluded that individuals express stronger emotional attachment with natural tourism settings than those with lesser experience. This in turn generates a sense of psychological well-being for visitors (Kaplan and Talbot, 1983; Korpela *et al.*, 2009). Some evidence suggests a significant positive relationship between place affect and pro-environmental behavioural intentions (Kals *et al.*, 1999; Ramkissoon *et al.*, 2013a). Place affect was an important predictor of attitudes towards preserving native forests (Pooley and O'Conner, 2000). Place affect may exhibit a mediating influence on visitors' satisfaction and their pro-environmental behaviour. If visitor satisfaction is high, engagement in pro-environmental behaviours at a specific natural and cultural site may also be high as a result of place affect. Studies exploring the mediating effect of place affect in such settings are scarce in literature. Accordingly, this study proposes the following:

> Proposition 1: Place affect mediates the relationship between place satisfaction and visitors' pro-environmental behaviour.

Place identity

Scholars have argued that place identity (Prohansky, 1978) is a well-established dimension of place attachment (e.g. Imran *et al.*, 2014; Ruiz and Hernandez, 2014). Place identity is defined as 'an individual's cognitions, beliefs, perceptions or thoughts that the self is invested in a particular spatial setting' (Jorgensen and Stedman, 2001, p. 238). Natural environments with distinctive features often foster higher levels of place identity in individuals (Budruk *et al.*, 2009; Halpenny, 2010; Ramkissoon *et al.*, 2013a). In the environmental psychology literature, researchers have suggested that continuous exposure to these settings generates a sense of identity to the place, leading to positive environmental behaviours (e.g. Bricker and Kerstetter, 2000; Hinds and Sparks, 2008; Walker and Ryan, 2008). Therefore, it is suggested that if levels of visitor satisfaction are high, the intentions to engage in pro-environmental behaviours as a result of place identity will also be high. The following is proposed with the aim of progressing literature on place identity, place satisfaction and pro-environmental behaviour:

> Proposition 2: Place identity mediates the relationship between place satisfaction and visitors' pro-environmental behaviour.

Place dependence

Individuals become attached to a place if it contributes to meeting their visitation goals (Williams *et al.*, 1992). Place dependence is defined as how best a place can provide desired leisure experiences (Williams *et al.*, 1992) and its relative quality compared to alternative places (Halpenny, 2010). This functional attachment emphasizes the importance of a resource in physical settings (Vaske and Kobrin, 2001) to cater for desired recreational activities such as hiking trails. Evidence shows indirect effects of place dependence on pro-environmental behaviour (e.g. Vaske and Kobrin, 2001). Ramkissoon *et al.* (2013a) found a significant positive effect of the former on the latter. No moderating effect of place dependence on visitors' pro-environmental behaviour was noted in Kyle *et al.*'s (2003) study. The inconclusive findings in literature suggest that the association between these constructs still need to be more explicitly documented. Place dependence may cause visitors to engage in pro-environmental behaviours through their satisfaction. The following proposition is based on the preceding discussion:

> Proposition 3: Place dependence mediates the relationship between place satisfaction and visitors' pro-environmental behaviour.

Place social bonding

Other studies have focused on another sub-dimension of place attachment, place social bonding. It is defined as the 'mutually constitutive' relationship between places and social relations (Wiles, 2005). Individuals may value a place if it provides them with social opportunities to redefine their 'being in place' (Hammitt, 2000; Scannell and Gifford, 2010). Natural settings set the context for social experiences, which, if maintained in these settings, are likely to lead to higher levels of attachment (Kyle *et al.*, 2004; Raymond *et al.*, 2010; Ramkissoon *et al.*, 2012) and foster group belongingness (Hammitt *et al.*, 2009). The argument that meanings of pro-environmental behaviours are often constructed through the social interaction of people in different settings (Georg, 1999; Nye and Hargreaves, 2009) has fuelled interest in exploring the people–place bond in natural settings (Ramkissoon *et al.*, 2013a). Place social bonding is likely to foster engagement in environmental actions in nature-based and cultural settings through visitor satisfaction. The following proposition emanates from the review of literature:

> Proposition 4: Place social bonding mediates the relationship between place satisfaction and visitors' pro-environmental behaviour.

7.2.3 Pro-environmental behaviour as a mediator

The quest for environmentally sustainable behaviour has led to an increased focus on human attitude and behaviour (e.g. Kurz, 2000; Ramkissoon and Uysal, 2014). Scholars are building on the attitude–behaviour model applying major psychological theories such as attachment theory (Bowlby, 1980b) and self-perception theory (Bem, 1972) to understand how place satisfaction contributes to an emotional attachment with environmental settings to encourage future visitation. Research on the mediating effect of pro-environmental behaviour in natural and cultural settings is scarce with the exception of a few studies (e.g. Ramkissoon and Mavondo, 2015). The latter study established that pro-environmental behaviour is one of the principal mechanisms linking place satisfaction and place attachment.

People may become attached to places that they are satisfied with and think will enhance their quality of life. Islands' distinctive attributes could lead to higher levels of place attachment enhancing the quality of life and well-being of visitors in a sustainable way. This is reinforced by Araujo de Azevedo *et al.* (2013), who argue that quality of life is enriched when people get the things they desire. Marketing places need to be aimed at fulfilling expectations of different market segments with the aim to promote quality of life (Kotler *et al.*, 2002). Parks and cultural attributes in island states not only provide health benefits as suggested in the health and well-being literature (e.g. Cohen *et al.*, 2007) but also promote social ties (Kimpton *et al.*, 2014; Ramkissoon and Mavondo, 2015), encouraging community sociability and belongingness.

This study proposes that pro-environmental behaviour mediates the relationship between place attachment and quality of life. Individuals with high levels of place attachment (affect, identity, dependence, social bonding) may engage in conscious efforts to protect an island's distinctive attractions. This may in turn enhance their quality of life. The following is proposed (Fig. 7.1):

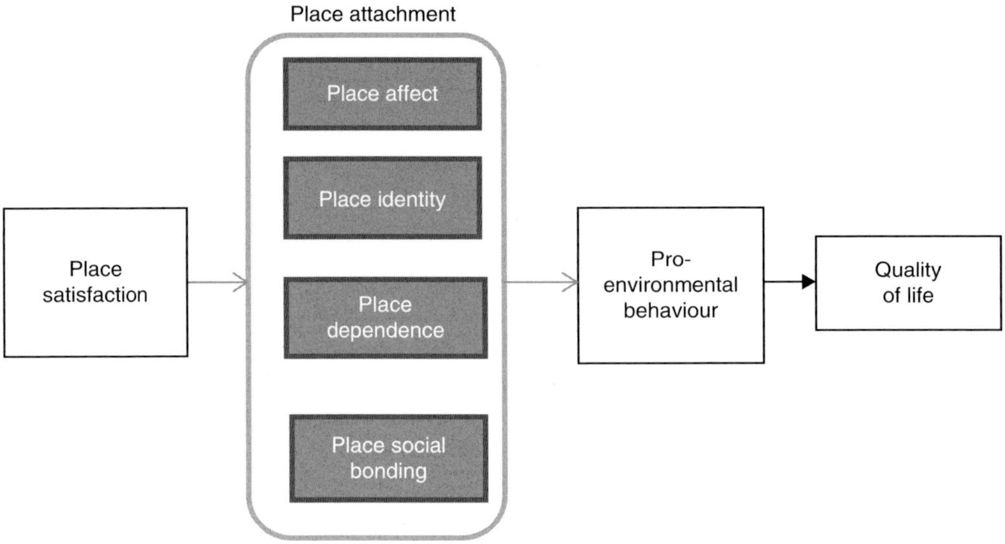

Fig. 7.1. Proposed conceptual model.

> Proposition 5: Pro-environmental behaviour mediates the relationship between (a) place affect and quality of life; (b) place identity and quality of life; (c) place dependence and quality of life; (d) place social bonding and quality of life.

7.3 Conclusions

This chapter can be perceived as the first step in the development of a conceptual framework contributing to our understanding of how visitor experiences are shaped by environmental psychological constructs. It develops and proposes a conceptual framework for island destinations, extending the theoretical interrelationships between local visitors' place satisfaction, place attachment (affect, identity, dependence, social bonding), pro-environmental behaviour and their overall quality of life in island economies. The conceptual framework provides a solid foundation with the hope that it will ascertain sustainable tourism development and enrich residents' overall quality of life in island destinations. To further develop and promote the tourism sector in island destinations, researchers could consider other elements of destination marketing and visitor experiences such as authenticity, motivation and fun. Incorporating these constructs in a single theoretical model may provide a deeper understanding of the interplay of factors influencing residents' quality of life in island and micro states. In order to test such complex models, one may need to use appropriate techniques such as structural equation modelling, path analysis and regression, which could be augmented with qualitative data obtained from local residents and other stakeholders in island destinations. Empirical testing of the conceptual framework will assist destination marketers and policy makers to enhance their marketing strategies not only for economic benefits, but also as a means for community development in island destinations (Nunkoo and Ramkissoon, 2010).

Destination marketers could develop and formulate effective strategies to foster higher levels of place attachment and pro-environmental behaviour. This will allow the local visitors to develop deeper emotional bonds with the natural and cultural settings, which may enhance their quality of life.

References

Araujo de Azevedo, A., Custodio, M. and Perna, F. (2013) 'Are you happy here?': The relationship between quality of life and place attachment. *Journal of Place Management and Development* 6(2), 102–119.

Baral, N., Stern, M. and Bhattarai, R. (2008) Contingent valuation of ecotourism in Annapurna Conservation Area, Nepal: Implications for sustainable park finance and local development. *Ecological Economics* 66(2/3), 218–227.

Bem, D. (1972) Constructing cross-situational consistencies in behavior: Some thoughts on Alker's critique of Mischel. *Journal of Personality* 40(1), 17–26.

Bowlby, J. (1980a) *Attachment and Loss: Loss,* Vol. 2. Basic Books, New York.

Bowlby, J. (1980b) *Attachment and Loss: Sadness and depression,* Vol. 3. Basic Books, New York.

Brehm, J., Eisenhauer, B. and Krannich, R. (2006) Community attachment as predictors of local environmental concern, the case for multiple dimensions of attachment. *American Behavioral Scientist* 50(2), 142–165.

Bricker, K.S. and Kerstetter, D.L. (2000) Level of specialization and place attachment: An exploratory study of whitewater recreationists. *Leisure Sciences* 22, 233–257.

Budruk, M., Thomas, H. and Tyrrell, T. (2009) Urban green spaces: A study of place attachment and environmental attitudes in India. *Society and Natural Resources* 22(9), 824–839.

Buhalis, D. (2000) Marketing the competitive destination of the future. *Tourism Management* 21, 97–116.

Cheng, T., Wu, H. and Huang, L. (2013) The influence of place attachment on the relationship between destination attractivess and environmentally responsible behaviour for island tourism in Penghu, Taiwan. *Journal of Sustainable Tourism* 21(8), 1166–1187.

Cohen, D., Mckenzie, T., Sehgal, A., Williamson, S., Golinelli, D. and Lurie, N. (2007) Contribution of public parks to physical activity. *American Journal of Public Health* 97(3), 509.

Crilley, G., Weber, D. and Taplin, R. (2012) Predicting visitor satisfaction in parks: Comparing the value of personal benefit attainment and service levels in Kakadu National Park, Australia. *Visitor Studies* 15(2), 217–237.

Croes, R. (2006) A paradigm shift to a new strategy for small island economies: embracing demand side economics for value enhancement and long term economic stability. *Tourism Management* 27, 453–465.

Davis, J., Le, B. and Coy, A. (2011) Building a model of commitment to the natural environment to predict ecological behavior and willingness to sacrifice. *Journal of Environmental Psychology* 31, 257–265.

Georg, S. (1999) The social shaping of household consumption. *Ecological Economics* 28(3), 455–466.

Gossling, S. (2003) *Tourism and Development in Tropical Islands: Political ecology perspectives.* Edward Elgar, Cheltenham, UK.

Halpenny, E. (2006) Environmental behaviour, place attachment and park visitation: A case study of visitors to Point Pelee national park. Unpublished doctoral thesis, University of Waterloo, Waterloo, Ontario.

Halpenny, E. (2010) Pro-environmental behaviors and park visitors: The effect of place attachment. *Journal of Environmental Psychology* 30(4), 409–421.

Hammitt, W.E. (2000) The relation between being away and privacy in urban forests recreation environments. *Environment and Behaviour* 32, 521–540.

Hammitt, W.E., Kyle, G.T. and Oh, C.O. (2009) Comparison of place bonding models in recreation resource management. *Journal of Leisure Research* 41(1), 57–72.
Hidalgo, M.C. and Hernandez, B. (2001) Place attachment: Conceptual and empirical questions. *Journal of Environmental Psychology* 21, 310–319.
Hinds, J. and Sparks, P. (2008) Engaging with the natural environment: The role of affective connection and identity. *Journal of Environmental Psychology* 28, 109–120.
Hoti, S., McAleer, M. and Shereef, R. (2007) Modelling international tourism and country risk spillovers for Cyprus and Malta. *Tourism Management* 28(6), 1472–1484.
Imran, S., Alam, K. and Beaumont, N. (2014) Environmental orientations and environmental behaviour. Perceptions of protected area tourism stakeholders. *Tourism Management* 40, 290–299.
Jorgensen, B. and Stedman, R. (2001) Sense of place as an attitude: Lakeshore owners' attitudes toward their properties. *Journal of Environmental Psychology* 21, 233–248.
Kals, E., Shumaker, D. and Montada, L. (1999) Emotional affinity toward nature as a motivational basis to protect nature. *Environment and Behavior* 31, 178–202.
Kaplan, S. and Talbot, J.F. (1983) Psychological benefits of a wilderness experience. In: Altman, I. and Wohlwill, J.F. (eds) *Human Behavior and the Environment: Advances in Theory and Research. Behavior and the Natural Environment,* Vol. 6. Plenum Press, New York, pp. 163–203.
Kelly, G. and Hosking, K. (2008) Nonpermanent residents, place attachment and 'sea change' communities. *Environment and Behavior* 40(4), 575–594.
Kim, H., Woo, E. and Uysal, M. (2015) Tourism experience and quality of life among elderly tourists. *Tourism Management* 46, 465–476.
Kimpton, A., Wickes, R. and Corcoran, J. (2014) Greenspace and place attachment: Do greener suburbs lead to greater residential place attachment. *Urban Policy and Research* 32(4), 477–497.
Korpela, K., Ylen, M., Tyrvainen, L. and Silvennoinen, H. (2009) Stability of self-reported favourite places and place attachment over a 10-month period. *Journal of Environmental Psychology* 29, 95–100.
Kotler, P., Hamlin, M., Rein, I. and Haider, D. (2002) *Marketing Asian Places.* Wiley, Singapore.
Kurz, T. (2000) The psychology of environmentally sustainable behaviour: Fitting together pieces of the puzzle. *Analyses of Social Issues and Public Policy* 2(1), 257–278.
Kyle, G.T., Absher, J.D. and Graefe, A.R. (2003) The moderating role of place attachment on the relationship between attitudes towards fees and spending preferences. *Leisure Sciences* 25, 1–18.
Kyle, G., Mowen, A. and Tarrant, M. (2004) Linking place preferences with place meaning: an examination of the relationship between place motivation and place attachment. *Journal of Environmental Psychology* 24, 439–454.
Lee, J., Kyle, G. and Scott, D. (2012) The mediating effect of place attachment on the relationship between festival satisfaction and loyalty to the festival hosting destination. *Journal of Travel Research* 51(6), 754–767.
Lee, M., Yang, H., Chen, Y. and Chen, T. (2010) A study of island tourism with global and local perspective. *Journal of Island Tourism Research* 3(1), 1–15.
Lopez-Mosquera, N. and Sanchez, M. (2011) Emotional and satisfaction benefits to visitors as explanatory factors in the monetary evaluation of environmental goods. An application to periurban green spaces. *Land Use Policy* 28, 151–166.
Lovelock, C. and Wirthz, J. (2007) *Services Marketing: People, technology, strategy*, 6th edn. Prentice Hall, New Jersey.
Mannell, R. and Kleiber, D. (1997) *A Social Psychology of Leisure*. Venture, State College, Pennsylvania.
McKercher, B., Wang, D. and Park, E. (2015) Social impacts as a function of place change. *Annals of Tourism Research* 50, 52–66.
Nunkoo, R. and Ramkissoon, H. (2010) Developing a community support model for tourism. *Annals of Tourism Research* 38(3), 964–988.
Nye, M. and Hargreaves, T. (2009) Exploring the social dynamics of proenvironmental behaviour change: A comparative study of intervention processes at home and work. *Journal of Industrial Ecology* 14(1), 137–149.

Paille, P. and Mejia-Morelos, J. (2014) Antecedents of pro-environmental behaviors at work: The moderating influence of psychological contract breach. *Journal of Environmental Psychology* 38, 124–131.

Pooley, J.A. and O'Conner, M. (2000) Environmental education and attitudes. *Environment and Behavior* 32, 711–723.

Prohansky, H.M. (1978) The city and self-identity. *Environment and Behavior* 10, 147–169.

Rahul, T. and Majhi, R. (2014) An adaptive non-linear approach for estimation of consumer satisfaction and loyalty in mobile phone sector in India. *Journal of Retailing and Consumer Services* 21, 570–580.

Ramkissoon, H. (2015) Authenticity, satisfaction, and place attachment: A conceptual framework for cultural tourism in African island economies. *Development Southern Africa* DOI: 10.1080/0376835X.2015.1010711.

Ramkissoon, H. and Mavondo, F. (2014) Pro-environmental behaviour: the link between place attachment and place satisfaction, *Tourism Analysis* 19(6), 673–688.

Ramkissoon, H. and Mavondo, F. (2015) The satisfaction–place attachment relationship: Potential mediators and moderators. *Journal of Business Research http:dx.doi.org.10.1016j.jbusres/2015.05.002*

Ramkissoon, H. and Uysal, M. (2010) Testing the role of authenticity in cultural tourism consumption: A case of Mauritius. *Tourism Analysis* 15, 571–583.

Ramkissoon, H. and Uysal, M. (2011) The effects of perceived authenticity, information search behavior, motivation and destination imagery on cultural behavioral intentions of tourists. *Current Issues in Tourism* 14(6), 537–562.

Ramkissoon, H. and Uysal, M.S. (2014) Authenticity as a value co-creator of tourism experiences. In: Prebensen, N.K., Chen, J.S. and Uysal, M. (eds) *Creating Experience Value in Tourism*. CAB International, Wallingford, UK, pp. 113–124.

Ramkissoon, H., Uysal M. and Brown, K. (2011) Relationship between destination image and behavioral intentions of tourists to consume cultural attractions. *Journal of Hospitality Marketing and Management* 20(5), 575–595.

Ramkissoon, H., Weiler, B. and Smith, L. (2012) Place attachment and pro-environmental behaviour in national parks: the development of a conceptual framework. *Journal of Sustainable Tourism* 20(2), 257–276.

Ramkissoon, H., Smith, L.G.D., and Weiler, B. (2013a) Relationship between place attachment, place satisfaction, and pro-environmental behaviour in an Australian national park. *Journal of Sustainable Tourism* 21(3), 434–457.

Ramkissoon, H., Smith, L.G.D. and Weiler, B. (2013b) Testing the dimensionality of place attachment and its relationships with place satisfaction and pro-environmental behaviours: A structural equation modelling approach. *Tourism Management* 36, 552–566.

Ramkissoon, H., Weiler, B. and Smith, L. (2013c) Place attachment, place satisfaction, and pro-environmental behaviour: A comparative assessment of place attachment and structural equation modelling. *Journal of Policy Research in Tourism Leisure, and Events* DOI:10.1080/19407963.2013.776371 (first published on 11 March 2013).

Ramkissoon, H., Smith, L. and Kneebone, S. (2014) Visitor satisfaction and place attachment in national parks. *Tourism Analysis* 19, 287–300.

Raymond, C.M., Brown, G. and Weber, D. (2010) The measurement of place attachment: Personal, community, and environmental connections. *Journal of Environmental Psychology* 30(4), 422–434.

Raymond, C., Brown, G. and Robinson, G. (2011) The influence of place attachment, and moral and normative concerns on the conservation of native vegetation: A test of two behavioural models. *Journal of Environmental Psychology* 31, 323–335.

Rodriguez, J., Parra-Lopez, E. and Yanes-Estevez, V. (2008) The sustainability of island destinations: Tourism area life cycle and teleological perspectives: The case of Tenerife. *Tourism Management* 29, 53–65.

Rolero, C. and De Picolli, N. (2010) Place attachment, identification and environment perception: An empirical study. *Journal of Environmental Psychology* 30, 198–205.

Ruiz, C. and Hernandez, B. (2014) Emotions and coping strategies during an episode of volcanic activity and their relations to place attachment. *Journal of Environmental Psychology* 38, 279–287.

Scannell, L. and Gifford, R. (2010) The relations between natural and civic place attachment and pro-environmental behavior. *Journal of Environmental Psychology* 30(3), 289–297.

Stedman, R. (2002) Toward a social psychology of place predicting behavior from place-based cognitions, attitude, and identity. *Environment and Behavior* 34(5), 561–581.

Stokols, D. and Shumacker, S.A. (1981) People in places: A transactional view of settings. In: Harvey, J. (ed.) *Cognition, Social Behavior, and the Environment*. Erlbaum, Hillsdale, New Jersey, pp. 441–448.

Tuan, Y. (1977) *Space and Place: The perspective of experience*. University of Minnesota Press, Minneapolis, Minnesota.

Tuan, Y. (1980) Rootedness versus sense of place. *Landscape* 24, 3–8.

Tudoran, A., Olsen, S. and Dopico, D. (2012) Satisfaction strength and intention to purchase a new product. *Journal of Consumer Behaviour* 11(5), 391–405.

Uysal, M. (2006) Factors of satisfaction: A case study of explorer parks. In: Prideaux, B., Moscardo, G. and Laws, E. (eds) *Managing Hospitality and Tourism Services*. CAB International, Wallingford, UK, pp. 350–357.

Uysal, M. and Williams, J. (2004) The role of expressive instrumental factors in measuring satisfaction. In: Grouch, G., Perdue, R., Timmermanns, H. and Uysal, M. (eds) *Consumer Psychology of Tourism, Hospitality and Leisure*. CAB International, Wallingford, UK, pp. 227–235.

Uzzell, D., Pol, E. and Badenas, D. (2002) Place identification, social cohesion and environmental sustainability. *Environment and Behavior* 34(1), 26–53.

Vaske, J. and Kobrin, K. (2001) Place attachment and environmentally responsible behavior. *Journal of Environmental Education* 32(4), 16–21.

Veasna, S., Wu, W. and Huang, C. (2013) The impact of destination source credibility: The mediating effect of destination attachment on destination image. *Tourism Management* 36, 511–526.

Venegas Sr, M. and Croes, R. (2003) Growth, development and tourism in a small economy: Evidence from Aruba. *International Journal of Tourism Research* 5, 315–330.

Walker, A.J. and Ryan, R. (2008) Place attachment and landscape preservation in rural New England: a Maine case study. *Landscape and Urban Planning* 86(2), 141–152.

Walker, G.J. and Chapman, R. (2003) Thinking like a park: The effects of sense of place, perspective-taking, and empathy on pro-environmental intentions. *Journal of Park and Recreation Administration* 21(4), 71–86.

Wiles, J. (2005) Conceptualizing place in the care of older people: the contributions of geographical gerontology. *Journal of Clinical Nursing*, 14(8B), 100–108.

Williams, D.R., Patterson, M.E., Roggenbuck, J.W. and Watson, A.E. (1992) Beyond the commodity metaphor: Examining emotional and symbolic attachment to place. *Leisure Sciences* 14, 29–46.

Wynveen, C., Kyle, G. and Sutton, S. (2012) Natural areas visitors' place meaning and place attachment ascribed to a marine setting. *Journal of Environmental Psychology* 32(4), 287–296.

Yuksel, A., Yuksel, F. and Bilim, Y. (2010) Destination attachment: Effects on customer satisfaction and cognitive, affective and conative loyalty. *Tourism Management* 31, 274–284.

Part III Sustainability and Alternative Resource Use

8 Developing a Sustainable Caribbean Tourism Product

Allan Wright*

Central Bank of Barbados, Caribbean Centre for Money & Finance and CEMLA

8.1 Introduction

There are various definitions of sustainability. Lowitt and Grimsley (2009) define it as encompassing environmental and social responsibility. According to Natural Capitalism Solutions, sustainability is the cautious and efficient protection of resources by businesses, communities and citizens. It is the practice of satisfying our needs in manners that take into account future generations and that invigorate natural and cultural assets.

Tourism sustainability has a more specific definition. The Caribbean Tourism Organisation defines sustainable tourism using the definition from the OECDS:

> Sustainable tourism is the optimal use of natural and cultural resources for national development on an equitable and self-sustaining basis to provide a unique visitor experience and an improved quality of life through partnerships amongst government, the private sector and communities.
>
> In accordance with this definition, the Bali Sustainable Development Project (BSDP) (1994) suggested that sustainable tourism should have the following features: '(i) the continuity of natural resources and production (ii) the continuity of culture and the balances within culture and (iii) development as the process which enhances the quality of life'.

Tourism's survival requires a balance between better product quality, profitability, inter-sectoral linkages, and mix between growth and taking care of the environment. Tourism leaders must also be creative when engaged in marketing and promotional campaigns to attract new customers and reduce the reliance on the North American and UK markets. They could focus on targeting emerging markets in Europe, Asia and Latin America, and even tourists in the Caribbean. According to Holder (2001), this is especially important in the context of a global economic slowdown, increasing

*E-mail: wright@cemla.org

airline fares and reduced air traffic. Holder (1996) states that the tourism sectors that involve a special interest, such as ecotourism, agro-tourism, adventure tourism, community tourism, health tourism, sports tourism and cultural tourism, need to be further analysed and subsequently incorporated as part of the sustainable product.

8.1.1 Core indicators of sustainable tourism

In order to help leaders determine if the tourism industry lies along a sustainable development path, one can utilize the WTO's (1995) 11 core indicators as guidelines for measuring sustainability. According to Greenidge and Greenidge (2011), these core indicators can be summarized into:

1. Tourism's contribution to the domestic economy.
2. Local community and tourist satisfaction levels.
3. Ecosystems.
4. Waste management.
5. Social impact and development control.
6. Facilities/site safety.

In 2004, these indicators were reviewed to make a more equitable distribution, and hence a new list was released which covered 12 indicators, including new guiding principles like tourism seasonality and developmental control. These new indicators are presented in Table 8.1, along with their explanations.

These indicators are a useful guide regarding how industries can achieve a sustainable competitive advantage in tourism. As a starting point, we examine the present state of the Caribbean tourism industry.

8.2 Caribbean Tourism Industry

8.2.1 Main aspects

The Small Island Developing States (SIDS) within the Caribbean have some common characteristics: they are small domestic markets, have no economies of scale and experience high transportation costs. This prevents them from undertaking development policies usually accessible to their larger counterparts. Throughout history, they have concentrated on exporting agriculture and, in a few cases, mining. However, in the last decades, they have undergone a significant decline in export agriculture (Hayle *et al.*, 2010). This, accompanied by their relatively undiversified production base, has forced many of them to seek alternative development policies, such as tourism promotion. Tourism promotion has been undertaken due to a perceived ability of this industry to create jobs and the potential multiplier impact of tourism spending, a perception that comes from these countries' natural sand, sea and sun endowments. Thus, tourism seemed like a natural industry to pursue and one in which they had a natural comparative advantage.

Over time, the tourism industry has become one of the main foreign exchange earners, a leading contributor of government revenue, and an important job provider in the Caribbean. Jamaica, the third main destination in the Caribbean, is the fifth most tourism-dependent country in the world with 2 million tourists in 2009 and its

Table 8.1. WTO (2004) sustainable tourism indicators.

	Issue	Indicator(s)
1	Community satisfaction with tourism	Community satisfaction level (measured by questionnaire)
2	Effect of tourism on local community	Tourist density ratio (tourist to locals) during off and peak season Percentage who believe that tourism has helped bring new services or infrastructure (questionnaire based) Number and capacity of social services available to community
3	Sustaining tourist satisfaction	Level of satisfaction by visitors (questionnaire) Percentage of return visitors Perception of value for money (questionnaire)
4	Tourism seasonality	Tourist arrivals by period Occupancy rates for licensed accommodation by period Percentage of business establishments open all year Number and percentage of tourism industry jobs which are permanent vs temporary
5	Economic benefits of tourism	Number of local people employed in tourism Revenues generated by tourism as percentage of total revenues generated in the community
6	Energy management	Per capita consumption of energy from all sources Percentage businesses participating in energy conservation programmes Percentage of energy consumption from renewable resources
7	Water availability and conservation	Water usage per tourist per day Water saved
8	Drinking water quality	Percentage of tourism establishments with water treated to international standards Frequency of water-borne diseases: number reported/ percentage of visitors
9	Sewage treatment – waste water management	Percentage of sewage from site receiving treatment Percentage of tourism establishments
10	Solid waste management	Waste volume produced by the destination by period Volume of waste recycled/total volume of waste Quantity of waste strewn in public areas
11	Development control	Existence of a land use or development planning process Percentage of area subject to control
12	Controlling use intensity	Total number of tourist arrivals Number of tourist per square metre of the site

tourism value-added contribution to GDP (CTO, 2010). In Antigua and Barbuda, tourism creates direct and indirect employment opportunities for approximately over 90% of its citizenry (WTO, 2009). In 2008, at least eight Caribbean states were listed within the top 30 nations worldwide as the most dependent on the tourism sector for direct or indirect employment. The high dependence on the sector for employment and foreign exchange demands a significant sectoral diversification to achieve more benefits (Hayle *et al.*, 2010).

However, increases in arrivals, employment and revenue have not caused improvements in the living standards of average Caribbean citizens, since approximately 36% of Latin America and the Caribbean's population live below the poverty line, the same percentage as a decade ago.

Therefore, one of the main challenges of governments in the Caribbean is to create and maintain a sustainable product that reduces indebtedness, promotes employment opportunities and raises the standard of living. However, before establishing how a sustainable product can be made, Caribbean tourist arrivals, receipts and the industry challenges must be reviewed.

8.2.2 Caribbean tourist arrivals and receipts (1999–2010)

In this section, a group decomposition analysis is done in order to compare tourist arrivals and receipts in the various countries. The groups are as follows:

1. The Bahamas, Barbados, the Cayman Islands and Jamaica (largest English-speaking destination markets);
2. Dominica, St Vincent and the Grenadines, St Lucia, Grenada (Windward Islands markets (A));
3. Barbados, St Lucia and Grenada (Windward Islands markets (B));
4. Anguilla, Antigua and Barbuda and Bermuda (Leeward Islands markets);
5. Aruba, Curacao and St. Maarten (Dutch Islands markets); and
6. The Dominican Republic and Puerto Rico (Spanish Islands markets).

All countries experienced a slowdown in arrivals and tourism receipts after the terrorist attacks of 11 September 2001 and after the global financial crisis of 2008 (after most countries had started to rebound after the terrorist attacks). Moreover, in some countries the arrivals and receipts had been negatively affected by hurricanes, particularly in groups (1) to (4). Hurricanes occurred in Grenada (2004 and 2006), St Lucia (2005), Antigua and Barbuda (2005) and in Bermuda (2005 and 2008), causing in some cases substantial damage to these markets' tourism infrastructure. For example, large parts of Grenada were declared a disaster zone.

Between these two shocks, it was a period of recovery and growth. Barbados and Anguilla benefited from the mentioned hurricanes. In Barbados, the hosting of the Cricket World Cup in 2007 aided its recovery. In Jamaica, government support for infrastructure and effective marketing techniques contributed to the formation of Spanish Hotel chains, which further promoted an increase of arrivals. At the end of 2009, Antigua and Barbuda stabilized by hosting a cricket T20 competition, which attracted international participation. Dominica had the fastest recovery after the terrorist attacks in 2001, due to a marketing campaign with an environmental focus in which nature-loving tourists were targeted. Saint Vincent and the Grenadines' recovery was slower relative to Dominica, but good marketing and increased investment in the Grenadines islands arrivals helped stabilize arrivals and tourism receipts.

In Aruba, after a decline in 2006, arrivals and receipts increased after a North American marketing campaign conducted after the suspected murder of an American teenager on the island. Saint Maarten experienced a stable increase in French and Dutch tourists since 2005. In Curacao, after 2004, duty free zones attracted regional business travellers, particularly those from the northern Caribbean and Trinidad.

In 2003, the Dominican Republic implemented a marketing campaign beyond North America, which included Asia, Latin America and Europe. As a result, there was a significant growth in arrivals and receipts in the Spanish market of the Dominican Republic. The increase of Spanish hoteliers and government support programmes to increase tourism infrastructure coincided with this promotional campaign. Puerto Rico, as it received the main visits from the USA, undertook efforts to expand its marketing segments in Latin America and Europe (particularly Spain) after 2004.

8.2.3 Main challenges

Although significant efforts have been made to expand output and employment in certain industries, such as agriculture (particularly bananas, sugar and coffee), oil refinery, petrochemicals, financial services and offshore banking, tourism remains the largest contributor to output, revenue and employment in the Caribbean (Craigwell, 2007). However, the CTO (2012) stated that in 2009 arrivals were down by about 800,000, 3.6% less than the previous year, up by 4.2% in 2010 and by 3.3% in 2011.

The general outlook is that the tourism growth rates in the Caribbean are low, and its international share remains quite low. Sharpley (2002) states that these markets have about US$63.8 billion of international receipts. The CTO (2010) identified as a great challenge for the tourism industry in the Caribbean the 'state of flux' of the global economy, which contributed to an overall weakening demand.

In addition to the challenges of achieving and maintaining high levels of tourist inflows and spending in a weak global economic environment, the tourism-intensive economies of the Caribbean are confronted with other issues that affect the tourism industry:

1. A low income multiplier, which implies the effects of tourism inflows on overall income in the Caribbean are small.
2. A high leakage rate, since a great portion of tourism revenues are amassed abroad.
3. A low competitive price index, which indicates that a market is unable to preserve its share relative to its competition.

Ramjeesingh (2008) calculates the tourism income multiplier for 18 SIDS, where 10 of these are Caribbean countries. In the Caribbean, two countries had multipliers below 1 (Grenada 0.72 and the Bahamas 0.89), while the other group ranged from 1.10 (Jamaica) to 1.39 (St Kitts and Nevis). The results are showed in Table 8.2.

Tourism spending brings revenue for any economy, and therefore it is expected to have an impact on the economy's output. The tourism income multiplier following the Keynesian calculation of Ramjeesingh *et al.* (2010a, b) is used to capture these effects, with a bigger multiplier indicating a larger effect of tourism on the economy. Table 8.3 shows other tourism-intensive SIDS, out of which four have similarly low

Table 8.2. The Tourism Income Multiplier in Selected Caribbean Countries (Ramjeesingh, 2008).

Country	Income multiplier
Antigua and Barbuda	1.18
The Bahamas	0.89
Barbados	1.27
Dominica	1.18
Dominican Republic	1.33
Grenada	0.72
Jamaica	1.10
St Kitts and Nevis	1.39
St Lucia	1.15
Trinidad and Tobago	1.32

Table 8.3. The Tourism Income Multiplier in selected tourism-dependent SIDS (Ramjeesingh, 2008).

Country	Income multiplier
Belize	1.01
Cape Verde	0.74
Comoros	1.14
Fiji Islands	0.66
Mauritius	1.03
Papua New Guinea	0.59
Seychelles	1.14
Singapore	0.71

multipliers (Cape Verde, Fiji Islands, Papua New Guinea and Singapore), with values below 1. Both Comoros and Seychelles have the highest multiplier of 1.14.

The high rates of leakage pose another formidable challenge to Caribbean tourism. Despite government policies to change the degree of import material in tourism products by increasing inter-sectoral linkages, the high rates remain. According to the World Travel and Tourism Council (WTTC, 2002) and the WTO (2004), the import of consumer and commercial goods, overseas marketing and profits repatriation greatly reduce net tourism earnings. Among the Caribbean states presented in Table 8.3, the highest leakage rates were in the Bahamas (0.85), Barbados (0.66) and St Lucia (0.62), while the lowest were recorded in Trinidad and Tobago (0.22) and Dominica (0.45). Table 8.4 depicts leakage rates in selected Caribbean markets.

The competitive index is formed from a mix of sub-indexes based on: (i) regulatory framework; (ii) business environment and infrastructure; and (iii) human, cultural and natural resources. According to the WTTC (2011), a higher index implies more competitiveness and greater ability to preserve its share relative to other

Table 8.4. Leakage rates in selected Caribbean markets (Ramjeesingh, 2006).

Country	Leakage rate
Antigua and Barbuda	0.56
The Bahamas	0.85
Barbados	0.66
Dominica	0.45
Dominican Republic	NA
Grenada	0.55
Jamaica	0.50
St Kitts and Nevis	0.60
St Lucia	0.62
Trinidad and Tobago	0.22

NA, not available

Table 8.5. The Tourism Competitive Index of selected Caribbean markets (WTTC, 2011).

Country	Rank	Overall index score
Barbados	29	4.79
Puerto Rico	45	4.42
Jamaica	60	4.12
Trinidad and Tobago	79	3.91
Guyana	98	3.62

nations. In analysing the index of major Caribbean countries, most of them have low indices, with the exception of Barbados which is the highest in Latin America and the Caribbean and second in the national perception towards the value of tourism. Barbados is currently ranked number 29 with an index of 4.79, followed by Puerto Rico at 45, Jamaica at 65, Trinidad and Tobago at 79 and Guyana at number 98. These rankings are presented in Table 8.5.

Hayle *et al.* (2010) state that other less recognized problems that afflict tourism are crime and harassment, whose cause has been signalled as a lack of community involvement. All of these factors, the linkage factors, low income multipliers, leakage rate and crime and harassment reduce the possibilities for achieving a sustainable competitive advantage (SCA) in the Caribbean tourism industry.

The Caribbean has highlighted expanded capacity for current and new tourist areas, but simultaneously the region has paid less attention to sustaining and improving product quality, engaging in promotion of the region, and in raising overall safety through robust environmental management systems. The Caribbean markets have not been able to find that balance between growth and sustainable development.

In essence, Caribbean states have failed to find the perfect fit between growth and sustainable development. In order to find this balance and be more competitive, the countries need to take advantage of ecotourism, agro-tourism and community tourism, and additionally complement these activities with inter-sectoral linkages to manufacturing, transportation, services, telecommunication and agriculture. They also need to improve product positioning to growing market segments which represent the highest potential demand (Holder, 1996, 2001; Jayawardena and McDavid, 2000). By doing so they would be able to capture a higher portion of the international market, hence creating a leverage to further improve the economic well-being of different stakeholders, including locals that are part of the tourism production system.

One possible way in which the challenges can be met and the industry could perhaps flourish is through the development of SCAs within the industry. Therefore, in the next section we analyse the main components of a competitive strategy and its relation to sustainability.

8.2.4 Competitive advantage

Day *et al.* (1997) use a model of competitive advantage, which explains that by investing in greater assets and different abilities, and expanding resources to new arenas, regional governments can develop a sustainable tourism industry. Figure 8.1 presents a graphical illustration of this process, called the competitive advantage cycle.

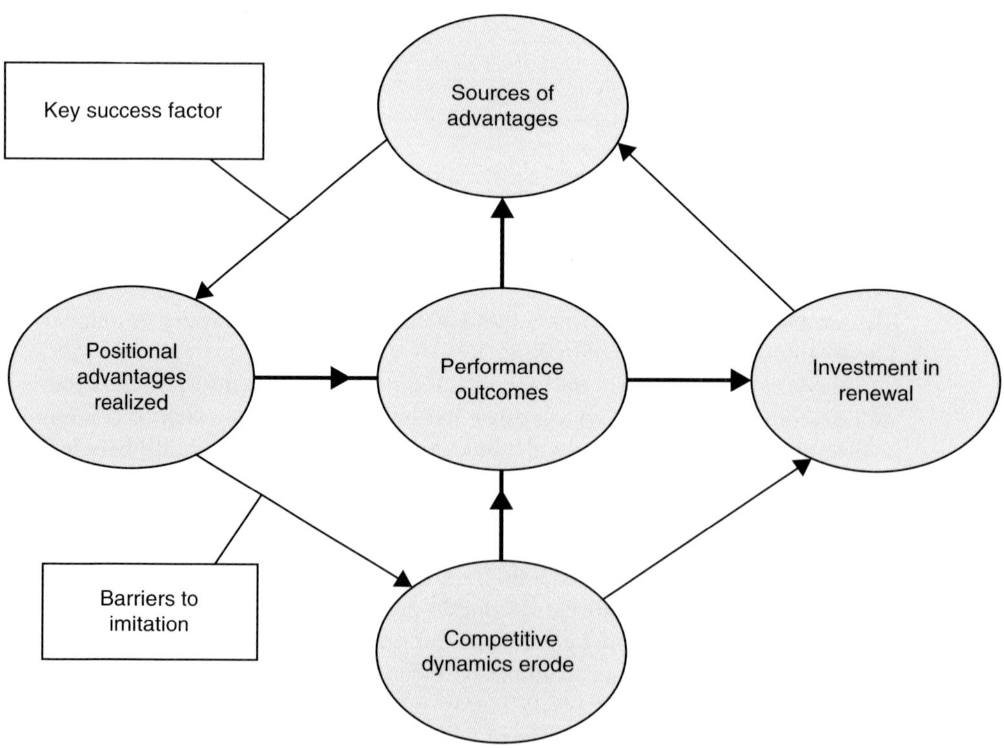

Fig. 8.1. Competitive advantage cycle (Day *et al.*, 1997).

To gain a competitive advantage within the industry will require a set of particular abilities and assets, such as expenditure in advertising, brand equity and room stock. The value of these assets is based on the good intentions constructed over the last years. According to Day *et al.* (1997) and Nustini (2009), competences consist in structural routines and best practices that are extremely hard to reproduce, which provide skills and information on the best ways of coordinating activities and using assets. This superiority in assets and abilities will contribute to defining the advantage one has and will encourage sustainable growth through increased profits, improved market share and better customer satisfaction. The unpredictable market dynamics and competitor's activities can easily wear away a region's sustainable position. Therefore, according to Luo (2002, 2004), industry leaders must promote reinvesting in regeneration processes and in activities that maintain the competitive edge within the region.

Other authors examine the notion of integrated economies to achieve a competitive advantage. Porter (1990, p. 119) recognizes how important it is to incorporate the concept of integrated economies in a sustainable strategy. They suggest that the cluster of firms in inter-related industries, institutions and business networks and relationships, can stimulate new ventures and enhance productivity. This particular approach of linkages through reinforced institutions is of crucial importance so that firms can remain sustainable and achieve a competitive advantage. Teo and Hiong (2003, p. 287) examine the notion of asset that linked economies achieve a faster transference of technologies and higher levels of local firm participation. This is achieved through the exports of important tourism products as well as higher levels of import substitution. By integrating the tourism-intensive economies, production could be stimulated through local firms adding higher value and improving job training within the sector.

8.2.5 Sustainable competitive advantage

In this section, we review the basics of competitive strategy and how it relates to sustainability, which leads us to the concept of sustainable competitive advantage (SCA). Porter (1985) established that the formulation of a competitive strategy requires relating the firm to its environment, and its goal is to find the firm's position in the industry where it can most efficiently and effectively deter competitive forces or can impact them in its favour. According to his classical model, cost and differentiation are two initial sources of competitive advantages. A low cost strategy is based on efficient and tight cost control, and differentiation implies offering an exclusive technology, brand, customer service and products to obtain increased market share. In general, Porter's model focuses on an organization's capabilities and competitive advantage, which originate in the firm's resources. This can be related to the sustainability indicators regarding the efficient management of infrastructure such as energy, sewerage and water systems, which are critical to achieving sustainability in the tourism industry.

This brings us to the two perspectives in the classical strategic literature: the resource-based view and capability-based view. The resource-based view establishes that firms achieve competitive advantage when they have valuable, uncommon and inimitable resources. The capability-based view argues that a firm's strategy success depends on the ability of transforming these resources into a competitive advantage, moving closer to understanding how organizations can sustain competitive advantage. Strategies through which competitive advantages can be obtained include improved quality, innovation, lower costs, improved processes and marketing.

However, the issue does not end at simply achieving a competitive advantage, but rather at making it sustainable (Warraich *et al.*, 2013). Competitive strategy and sustainability are related through a concept called SCA. The definition of SCA has evolved throughout the decades. According to Hofer and Schendel (1978), it is the unique position that an organization or firm develops relative to competitors that allows them to constantly outperform them. This happens when the firm has certain value-creating processes and positions that are not easily imitated by other firms. Therefore, it must be maintained for a great amount of time despite the competition's actions. This differs from competitive advantage, because it focuses on an advantage that is not easily replicated in the long term. This can be related to the sustainable indicators for the Caribbean industry as a whole; the Caribbean tourism industry must develop a unique position through an aggregate product that cannot be imitated by other regions that offer similar venues, which would lead to increased tourist satisfaction and therefore increased visitors.

Porter (1985) establishes that in the long run, performance that is above the average is fundamental to achieve an SCA. Iyiola (2010) states that firms can acquire SCA through human capital resource, physical capital resources, organizational capital resources and culture, and corporate entrepreneurship. Iyiola then provides a valuable analysis of the literature review surrounding SCA. Barney (1991) states that for something to be considered an SCA, it must take advantage of opportunities and defuse threats, be rare among the competition, not be easily imitable, and there cannot be tactically equal substitutes for it. In accordance, he establishes four indicators that signal the potential for a firm's resources: value, rareness, imitability and substitutability. According to Lowson (2002), SCA can be based on knowledge, experience, innovation, know-how, and unique information accumulated throughout time. Coplin (2002) argues that most academics have focused their studies on firm-specific unique features that add value and are transferable to a variety of industry contexts. Goh (2003) notes that many firms are establishing a system of continuous learning in order to remain competitive. It has been argued that multiple types of intellectual property were not considered by Barney, such as patents, copyrights and trademarks. The set of patents held by a firm is one of the major intellectual property resources of a firm, and therefore has the potential to create an SCA. Strand argues that patents create value and can neutralize external threats from outside competition. Supporting Strand's arguments, other authors have found that patents foster value creation through linkages to other resources, such as production, finance and marketing activities. Njuguna (2009) analyses the role of organizational learning in helping firms obtain an SCA through the development of knowledge resources and capabilities. In attempting to determine if organizational learning can lead to SCA, he develops a model that links organizational learning to SCA through intellectual capital elements. In the model, organizational learning, which involves the acquisition of knowledge, knowledge distribution and knowledge application, enhances intellectual capital (human, social and organizational capital), which ultimately leads to the possibility of acquiring an SCA. These three types of intellectual capital reinforce each other, leading to strong feedback effects.

8.2.6 Recent trends

A brief examination is now presented regarding how global firms interpret sustainability, and in some instances how they relate it to their individual competitive advantages. In a recent survey done by the Boson Consulting Group and MIT in 2012 for more than 4000 managers in 113 countries, almost one-third said that their sustainability activities are contributing to their profitability. This group is called the 'Harvesters', and they basically state that their sustainability activities contribute to their profits. Harvesters are not only lowering carbon emissions and decreasing their consumption of energy, they are also changing their operating frameworks and strategies. The report identified another group called the 'Embracers', who see sustainability as a requirement to being competitive. One of the reasons, according to the survey, is that customer preferences for sustainable products is increasing. Institutional investors are demanding more information on companies' sustainability performances. These investors more frequently interpret sustainability as an indicator of business value and profitability. Large companies find themselves compared in this area. Companies mentioned increased innovations as one of the benefits of sustainability practices. What Harvesters are doing is establishing new structures, lines of communication and performance metrics, rather than implanting new sustainability practices into previous organizational arrangements.

8.3 Strategies That Companies Have Employed to Develop a Sustainable Product

Some companies have followed some of the principles of Porter's classical model to attain competitive advantages. Wal-Mart has pursued a low cost strategy through a low cost of advertising, superior logistics, a flat organizational structure and efficiency in the use of capital (He, 2012). Southwest Airlines also pursues a low cost strategy by flying only point to point lines. In Spain, Zara has achieved differentiation in the design, information systems and logistics management of its products. Zara owns its factories, designers and stores, and their designers have frequent contact with the stores and can therefore respond quickly to changing trends. The effect is that Zara's products reach stores in about 5 weeks, when the products of most fashion companies take 9 months to reach the stores (He, 2012). However, in other cases, the initial competitive advantage may disappear quickly, especially when companies find it difficult to adapt to new conditions. This happened to IBM, which started to lose market status when it failed to create new products in a context of dynamic consumer demand, so it resorted to simply protecting the old product: mainframes and mainframe technology. IBM had become too committed to its core competitive advantage, and then was unable to change direction for a long time. After 1993, Louis Gerstner took control of the firm and started to promote efficiency and innovation, as well as a culture of customer satisfaction and service. The company developed an innovative business model focusing on high profit services, electronic business and other technological innovations. IBM automated business systems such as finance, procurement and human resources. Profitability has steadily increased and the company successfully reinvented itself, becoming an object of frequent study. Although IBM was finally able to adapt, other companies were not so lucky. In the 1990s, DEC Company had a competitive advantage in producing high quality VAX minicomputers. However, the company developed a very narrow and specialized engineering focus, ignoring other important functions such as services and marketing. The firm was unable to adapt to the dynamic industry trends and the company was acquired by Compaq in 1998. This implies that to have a sustainable competitive advantage, one must be adaptable to evolving circumstances, and in some instances even transform themselves. In this sense, the tourism industry in the Caribbean needs to learn to adapt to the global business economic cycles, since it is mainly dependent on the US and UK markets. By diversifying its tourism base, it would be more adaptable to the dynamic and turbulent circumstances of the global economy.

Apple provides another very good example of a transformational story, since it was transformed from a PC firm to a flexible and innovative corporation (He, 2012). After failing in its early years as computer manufacturer, Apple pursued a differentiation strategy, achieving and maintaining a competitive advantage through the provision of a unique and different products in a 'commodity' industry, which are easy to use, elegantly designed, and embodies brand loyalty. It has also diversified its business core by engaging in the personal music players, music retail, tablet computing and TV hardware manufacturing sectors (He, 2012). It has a dominant market position in mobile phones applications, and has achieved this by rewarding the developers of these applications by sharing in the profits. Apple realized the importance of applications for the sale of iPhones, and thus has created incentives for third-party developers to make and sell them. In this sense, Apple has been able to leverage a network effect – in which the value of a product or service grows as the number of users grows – in a successful manner: the more valuable apps offered, the greater the utility of having an iPhone.

Even in emerging markets, innovative approaches to sustainability are occurring, with a more socially cognizant approach. In 1998, CEMEX, Mexico's top cement producer, founded Patrimonio Hoy, a savings club aimed at low-income people who want to build their own houses. CEMEX receives weekly savings payments from these people, and in exchange provides them with the infrastructure to build their homes. The end result is a higher quality home produced faster and at a lower cost. This is an exemplary way in which infrastructure is provided at a low cost. Although one cannot simply apply this programme to the Caribbean tourism industry, they can certainly take into account the guiding principles of infrastructure provision at a low cost when they are formulating their strategies.

Some companies, like Shell, have instituted roles of central control and execution. Shell has a small global business team which drives this change, but the main execution takes place at a firm-wide level. Campbell Soup has four sustainability teams in areas such as the environment and the community. In these specialized teams is where you get the expertise and decision making, and it is where ultimate accountability resides. If no one has ultimate accountability for driving sustainability, then it will not work. In this sense, a good idea would be for each Caribbean country to assign a specialized team in order to develop an SCA tailored to its own circumstances. If no one is ultimately responsible for this, it is unlikely that an SCA will be achieved.

At Kimberly-Clark, advancement is measured in accordance with sustainability objectives. There are quarterly updates on data concerning waste generation, the sales of environmentally inventive products, and energy and water usage, which are shown to stakeholders. Sometimes, companies employ strategies in response to a changing environment, which lead to the development of sustainable and innovative practices. In response to changing preferences and fuel costs, in 2007 BMW established 'Project i' to explore new mobility technologies, in which management handpicked the best engineers throughout the company. This eventually led to the technology stage of the electric vehicle programme. The project created an innovative environment, which facilitates the birth of sustainable competitive practices. These companies have also become more collaborative with people outside the company, such as suppliers. This is especially important for multinational companies. For example, in 2010 Procter & Gamble launched a sustainability scorecard to assess how suppliers performed in areas like waste management, gas emissions, water usage and other areas, which also allows it to promote innovation. The topic of service quality has also gained ground in the study of SCA.

Warraich *et al.* (2013) analyse whether service quality is a source of competitive advantage for firms in the telecommunications sector of Pakistan, and also examines whether customer intentions can be used as an indicator of SCA. To achieve this, a survey was carried out targeting 400 current users of cell-phones. The study found that the telecommunications industry in Pakistan received good ratings on equipment, but low ratings on empathy, particularly in not knowing customer's needs and in not providing enough personal attention.

8.4 Conclusions

The Caribbean faces multiple challenges: a high leakage rate, low income multiplier (except for Barbados) and medium–low competitiveness. In addition, there are persistent social issues such as a high crime rate and tourist harassment. Despite these daunting problems, the opportunity to create a sustainable tourism product which is both economically feasible and environmentally friendly is present. According to Jayawardena (2002), if the Caribbean governments want to augment the region as a sustainable product, they need to promote these strategies by maintaining suitable economic and social levels, high quality safety nets, and promote local population participation. Jayawardena (2000) backs the notion of applying good environmental practices, customer service at par with international standards, and quality practices in the context of the region's uniqueness. The Caribbean Tourism Human Resource Council (2002) recommended a country differentiation plan, which recognizes the region's exceptionality in order to achieve sustainable growth outcomes.

In order to develop a sustainable and efficient tourism service in the Caribbean, certain infrastructure must be in place, such as an easy free flow of people, enhanced air-linkages and an adequate functioning of telecommunications, transports, public utilities, health care and educational and business services. A sustainable approach in tourism should do the following.

1. Foment the development of local industries through import substitution and reduce the leakage rate as new products are improved.
2. Augment the sector's contribution to gross domestic product and to the region's development through the solidification of sustainable development and environmental systems.
3. Include local communities and a greater involvement with citizenry to improve the accumulation of benefits.
4. Link regional policies and solutions to common objectives.
5. Offer inter-sectoral linkages through higher multiplied income effects and contribute to developing other sectors (e.g. telecommunications, services, transport, manufacturing and agriculture).

Therefore, a summary of possible approaches for the Caribbean include the following.

1. Provide a good balance of inter-sectoral linkages, product quality, profitability, and a solid mix between economic growth and the environment.
2. Back tourism strategies through effective marketing, which could include attracting new market segments such as emerging countries and intra-regional travellers, and reducing reliance on North American and UK tourists.
3. Fomenting the development of tourism sub-sectors of special interest, such as agro-tourism, local tourism, health tourism, ecotourism and sport and cultural tourism, to expand the product's scope.
4. Assuring improved airline linkages, the proper simplification of free movement of people, adequate functioning transport, public utilities, educational and business services, health care, business services and telecommunications.
5. Investment in greater assets and different abilities, as well as expanding investment in new competitive arenas.

References

Barney, J.B. (1991) Firm resources and sustained competitive advantage. *Journal of Management* 17(1), 99–120.

Caribbean Tourism Organisation Human Resource Council (2002) *Building a Competitive Caribbean Tourism Workforce, 2002-2004 Strategic Plan*. Bridgetown, Barbados.

Caribbean Tourism Organisation (2010) *Caribbean Tourism Statistical Report*. Bridgetown, Barbados.

Caribbean Tourism Organisation (2012) *Caribbean Tourism: State of Industry*. Bridgetown, Barbados.

Coplin, L.H. (2002) Competitive advantage and the SMEs: the role of distinctive competences as determinants of success. PhD Thesis, Universitat Autonoma de Barcelona.

Craigwell, R. (2007) Tourism competitiveness in small island developing states. *UNU-WIDER Research Paper* 2007/19, United Nations University, Helsinki, Finland.

Crespo, N. and Diaz, S.N. (1997) Cuban tourism in 2007: economic impact. In: *Cuba in Transition* 7. Association for the Study of the Cuban Economy, Florida.

Day, G., Reibstein, D.J. and Gunther, R.E. (1997) *Wharton on Dynamic Competitive Strategy*. John Wiley & Sons, Canada.

Goh, S.C. (2003) Improving organizational learning capability: lessons from two case studies. *The Learning Organization* 10, 4.

Greenidge, K. and Greenidge, N. (2011) *Sustainable Tourism Development: The Case of Barbados*. Central Bank of Barbados, Barbados.

Hayle, C., Ramjeesingh, D. and Wright, A. (2010) Planning for Tourism Resilience in the Caribbean. *Journal of Eastern Caribbean Studies* 35(1), 36–59.

He, N. (2012) How to maintain sustainable competitive advantages – case study on the evolution of organizational strategic management. *International Journal of Business Administration* 3, 5.

Hofer, C. and Schendel, D. (1978) *Strategy Formulation: Analytical Concepts*. West, St Paul, Minnesota.

Holder, J. (1996) Marketing competitiveness in a new world order: regional solutions to Caribbean tourism sustainability problems. In: Harrison, L.C. and Husbands, W. (eds) *Practising Responsible Tourism*. Wiley, Toronto, pp. 145–173.

Holder, J. (2001) *Some Ideas for Tourism Change in a Changing World*. Caribbean Tourism Organisation, Barbados.

Iyiola, O. (2010) *Sustainability of Competitive Advantage: A Must for Every Firms*. College of Business and Social Studies of Covenant University, Nigeria.

Jayawardena, C. (2000) An analysis of tourism in the Caribbean. *Worldwide Hospitality and Tourism Trends* 1, 3.

Jayawardena, C. (2002) Cuba: crown princess of Caribbean tourism? Paper presented at the 27th Caribbean Association Conference, Nassau, the Bahamas.

Jayawardena, C. and McDavid, H. (2000) *Implications of Agro-Tourism – A Sustainable Approach to Economic Growth*. Inter-American Institute for Co-operation on Agriculture, Trinidad and Tobago, pp. 119–135.

Lowitt E.M. and Grimsley J. (2009) *Hewlett-Packard: Sustainability as a Competitive Advantage*. Accenture Institute for High Performance.

Lowson, R.H. (2002) *Strategic Operations Management: The New Competitive Advantage*. Routledge, London

Luo, Y.D. (2002) Organisational dynamics and global integration: A perspective from subsidiary managers. *Journal of International Management* 8(2), 189–215.

Luo, Y.D. (2004) *Co-opetition in International Business*. Copenhagen Business School Press, Copenhagen.

Njuguna, J. (2009) Strategic positioning for sustainable competitive advantage: an organizational learning approach. *KCA Journal of Business Management* 2(1), 32–43.

Nustini, Y. (2009) Determinants of sustainable competitive advantage due to an IT-enabled strategy. *Journal SlasatBisnis* 11, 1.

Porter, M. (1985) *Competitive Advantage: Creating and Sustaining Superior Performance*. Free Press, New York.

Porter, M. (1990) *The Competitive Advantage of Nations*. Macmillan Press, London.

Ramjeesingh, D. (2008) Small island developing states (SIDS): tourism and economic development. *Tourism Analysis* 13(5), 629–636.

Ramjeesingh, D., Wright, A. and Hayle, C. (2010a) Factors influencing hotel occupancy in Jamaica: Role of Events. *Tourism Analysis* 15, 3.

Ramjeesingh, D., Wright, A., Hayle, C. and Craigwell, R. (2010b) Is the tourism-led growth thesis valid? The case of Bahamas, Barbados and Jamaica. *Tourism Analysis* 15, 4.

Sharpley, R. (2002) The consumption of tourism. In: Sharpley, R. and Telfer, D.J. (eds) *Tourism and Development: Concepts and Issues*. Channel View Publications, Clevedon, UK, p. 300.

Teo, P. and Hiong, L. (2003) Global and local interaction in tourism. *Annals of Tourism Research* 30, 2.

Warraich, K., Warraich, I. and Asif, M. (2013) Achieving sustainable competitive advantage through service quality: An analysis of Pakistan's telecom sector. *Global Journal of Management and Business Research* 13, 2.

World Tourism Organization (1995) *What Tourism Managers Need to Know: A Practical Guide to the Development and Use of Indicators of Sustainable Tourism*. World Tourism Organization, Madrid, Spain.

World Tourism Organization (2004) *Indicators of Sustainable Development for Tourism Destinations: A Guidebook.* World Tourism Organization, Madrid, Spain.

World Tourism Organization (WTO) (2009) *International Tourist Arrivals.* Bank of Jamaica: Annual Statistical Digest, Kingston, Jamaica.

WTTC (World Travel and Tourism Council) reports commissioned and policy reports and initiatives (2002) Available at: http:// en.wikipedia.org/wiki/world_travel_and_tourism_council (accessed 4 March 2014).

WTTC (World Travel and Tourism Council) reports commissioned and policy reports and initiatives (2011) Available at: http://en.wikipedia.org/wiki/world_travel_and_tourism_council (accessed 4 March 2014).

9 Dive Tourism and the Entrepreneurial Process in the Perhentian Islands, Malaysia

Julia Jeyacheya[1]* and Mark P. Hampton[2]

[1]University of Bradford, UK; [2]University of Kent, UK

9.1 Introduction

Dive tourism is a high growth, niche sector for Small Island Developing States (SIDS) and countries with long coastlines and is propelled predominantly by local entrepreneurs and small businesses. This chapter examines dive tourism in a small Malaysian island, particularly factors influencing the entrepreneurial process.

Tourism is typically dominated by small or medium sized enterprises and creates many opportunities for entrepreneurial development and employment both directly (e.g. craftwork, retail units) and indirectly (agricultural producers) linked to tourism (Hampton and Jeyacheya, 2013). Although in principle there are many opportunities for entrepreneurs, the reality for many local communities is less clear because factors crucial for entrepreneurial success (loans, access to supply chains, skills training) are not readily available, accessible or, sometimes, unnecessary for business operations (Jaafar *et al.*, 2011). In the context of the scuba diving industry, Dimmock *et al.* (2013: 165) proposed four key components crucial for its survival (equipment, education, experience and environment) and thus the survival of the dive tourism sector. However, the reality for some communities in less developed countries (LDCs) is somewhat different.

Despite these obstacles, tourism entrepreneurs contribute significantly to tourism development (Ryan *et al.*, 2012) in many coastal and island destinations, including Malaysia. This suggests that the entrepreneurial process is influenced by a range of variables including social, cultural, political, geographical and historical (Bygrave, 2004; Ariff and Abubakar, 2005; Dana *et al.*, 2014).

Malaysia hosts a sizeable tourism industry with 27.4 million international arrivals in 2014 (UN WTO, 2015) and tourism contributed around 6% of GDP and directly employed 5.3% of the labour force (WTTC, 2015). The Ministry of Tourism

*E-mail: j.jeyacheya@bradford.ac.uk

and Culture views dive tourism as requiring minimal investment in comparison to resorts and other water sports, having high growth rates, being lucrative, high yield and with a broadly sustainable market (Basiron, 1994; Tourism Malaysia, 2009). As a result, dive training programmes have been promoted by the federal government since the 1990s to encourage local Malays (*Bumiputra*) to enter the industry (Basiron, 1994). This additional government support was part of a broader remit to encourage entrepreneurship, particularly for the *Bumiputra*, through increased funding, business advisory services and higher education.

This chapter examines the dive tourism entrepreneurial process in the Perhentian islands – one of Malaysia's most significant dive locations located about 20 km off the east coast of the peninsula (Fig. 9.1). We explore the question of *how* the tourism entrepreneurial process transforms (or not) as the destination moves to a strategically competitive one, using a holistic interpretation of the critical factors influencing change.

9.2 The Entrepreneurial Process in Tourism

Entrepreneurship has become a complex, widely applied concept within different research disciplines and is defined within the social and cultural norms of a given society (Gartner, 1990; Ariff and Abubakar, 2005). Of the many existing definitions, the individual is more often studied than the *process* itself (Morrison *et al.*, 1999).

Tourism entrepreneurship has been broadly defined by a range of factors related to individuals' motivation, economic status, background, social circumstances, personality characteristics and traits (Morrison, 2006; Jaafar *et al.*, 2011). Research examines a variety of personal, sociological and environmental factors associated with individual entrepreneurs and this suggests significant heterogeneity (Bjerke, 2000) between different tourism sectors (Koh and Hatten, 2002). Heterogeneity is also observed between local and in-migrant entrepreneurs (Dana *et al.*, 2014) and between developed countries and LDCs. Thus, personal attributes typically associated with entrepreneurs such as risk-taking, independence, innovation, self-confidence, a need for achievement (Jaafar *et al.*, 2011: 829; Lee *et al.*, 2011) and being proactive (Morris and Lewis, 1991: 23) are not commonly recognized in tourism entrepreneurs, particularly those operating small-scale firms. Tourism entrepreneurs are generally driven by lifestyle factors not profit; have a tendency to 'stay within the fence' by 'consciously reject[ing] certain economic and business growth opportunities' (Ateljevic and Doorne, 2000) and business advice from others (Hollick and Braun, 2005). Similarly, LDC tourism entrepreneurs appear less inclined towards profit maximization (DeBerry-Spence and Elliot, 2012; Ahmad *et al.*, 2014), are less likely to participate in 'risk-laden activities' (Jaafar *et al.*, 2011), and limit creativity and innovation (Kodithuwakku and Rosa, 2002).

This brief list broadly categorizes the tourism entrepreneur as a 'lifestyle entrepreneur' who pursues a 'desired lifestyle ... with little/no intention of growing their enterprises' (Koh and Hatten, 2002: 36) and who seemingly take little risk. However, given the geographical range of much research and the diversity of socio-economic groups, other factors embedded from the norms and circumstances of individual societies shape entrepreneurs' decisions and therefore the entrepreneurial process (DeBerry-Spence and Elliot, 2012: 1672). Also, common perceptions of tourism being an industry with low entry barriers where few skills are required and limited

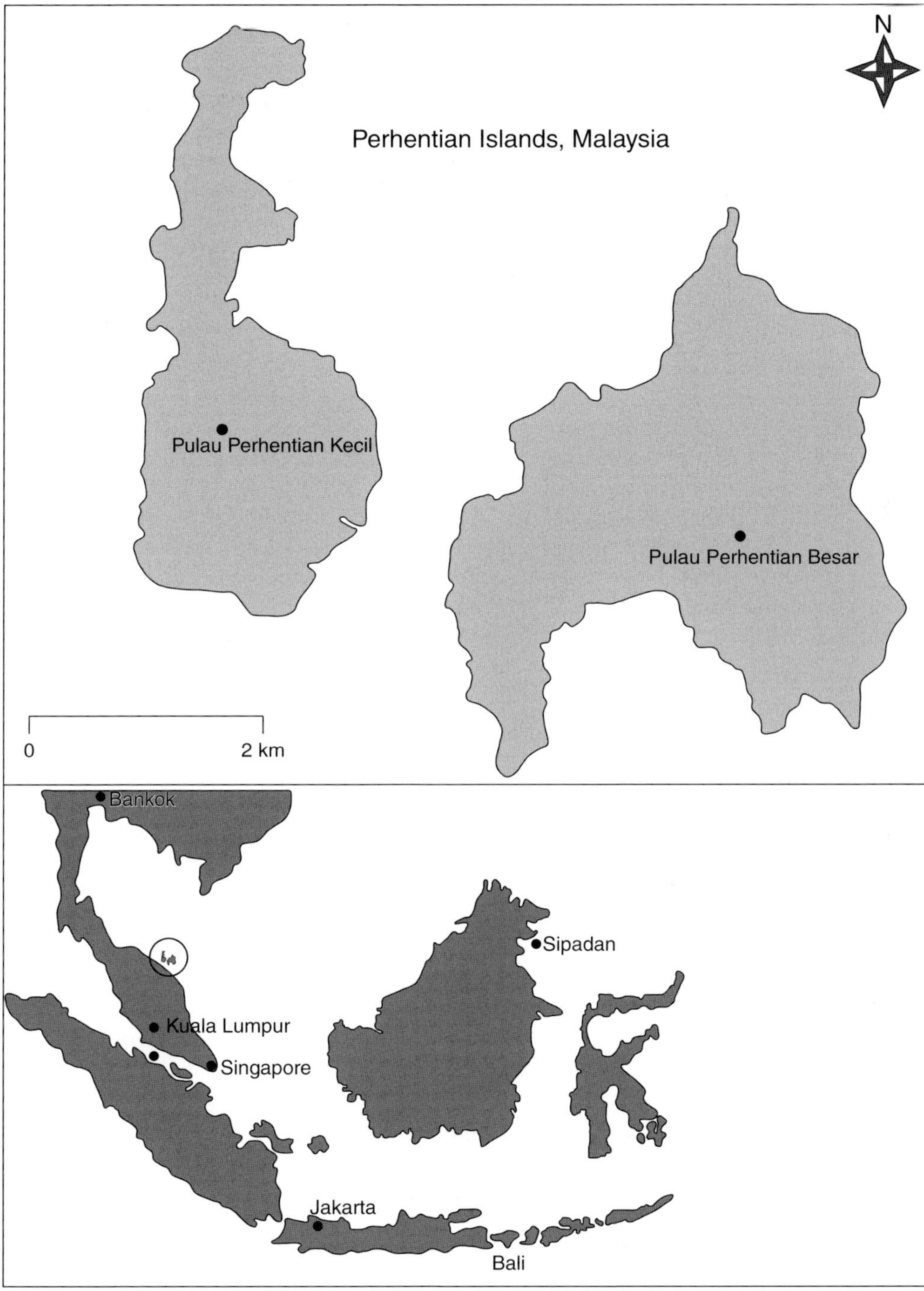

Fig. 9.1. Map of Perhentian Islands.

training is available (Hollick and Braun, 2005) may partially answer why it appears to attract a predominantly 'lifestyle'-led approach. This has led to the dominance of small and medium-sized tourism enterprises with many being linked to the informal sector (Scheyvens, 2006; Mograbi and Rogerson, 2007; Jaafar *et al.*, 2011) and in some cases, choose to imitate rather than innovate (Koh and Hatten, 2002).

None the less, tourism entrepreneurs are critical to the sustainable socio-economic development of their community (Morris and Lewis, 1991). Timmons (1999) contends they are adaptive and flexible to change, and such dynamism stimulates community involvement in tourism. McKercher (1999: 433) goes further, dubbing tourism entrepreneurs 'rogues or chaos makers' with the ability to 'single-handedly transform an organisation, destination or region or country.'

Tourism entrepreneurs have the capacity to instigate and sustain entrepreneurial activity (and thus the process) within their communities, raising the fundamental question of *how* this is achieved. Although much research examines the entrepreneurial process, the models are somewhat fragmentary and the process is conceptualized from inconsistent theoretical perspectives creating a knowledge gap of the practical implications.

This chapter does not develop a process model, but explores the question of *how* the tourism entrepreneurial process transforms (or not) in the Perhentian islands as an example of a small tourism-dependent island economy. It takes a holistic approach by considering the personal (characteristics, traits and socio-economic factors), environmental (business opportunities, resources and threats) and sociological factors (cultural norms, social networks, family) to understand how they interact with each other, with each stage of the process, and what the practical implications are. In this chapter, the process stages are assumed to be dynamic and influenced by past and present 'human volition' (Bygrave, 1993), and they follow a distinct path including the discovery of opportunity and/or the innovation of an idea, a triggering event, the implementation of the business and its growth.

9.3 Dive Tourism: The Literature

Within international tourism, dive tourism is a fast growing segment but little robust data exist concerning its size. PADI (Professional Association of Diving Instructors) – the largest certifying agency – estimates 3–6 million divers worldwide whereas safety organization DAN (Divers Alert Network) quotes (from accident reports) 1–3 million divers worldwide but gives a lower estimate of 'active divers' (more than five dives per year) of around 1–1.4 million (Davison, 2007). Regarding initial training and other qualifications, PADI certified 936,149 divers worldwide in 2013, bringing the cumulative total of divers to 22.2 million (1967–2013) (PADI, 2014). However, no international agreed definition exists, but it has been usefully (but broadly) defined by Tourism Queensland (2003: 1) as 'travel where at least one scuba diving expedition is included.' Nevertheless, for host regions dive tourists are an important market segment as they are typically from higher socio-economic groups and are high-spend visitors (Garrod, 2008). Unsurprisingly, dive tourism is seen as a key growth sector by tourism planners in many LDCs such as Jamaica, the Maldives etc. (Commonwealth Secretariat, 2002; Ministry of Tourism, Arts and Culture, 2012). Many LDCs, especially SIDS or those with long coastlines, already prioritize marine tourism (including so-called

'3S' tourism: sun, sea and sand) so that dive tourism appears a logical development utilizing marine resources to further grow income and employment.

Most of the dive tourism literature originates from the physical sciences and concentrates on environmental aspects, divers' biophysical impacts and reef conservation (e.g. Clifton, 2004; Worachananant *et al.*, 2008). In the social sciences there is a small but growing literature on dive tourism. Research examined: diver characteristics and motivations (Musa, 2002, 2003; Edney, 2012); diver typologies and behaviours (Garrod, 2008; Ong and Musa, 2011); the scuba diving tourism system (Dimmock and Musa, 2015); social and economic impacts of dive tourism (in Honduras – Canty, 2007; South Africa – Mograbi and Rogerson, 2007; Malaysia – Daldeniz and Hampton, 2013; Indonesia – Davis, 2005; Klimmek, 2013); stakeholders and sustainability (Haddock-Fraser and Hampton, 2012) and dive tourism using resilience theory (Hillmer-Pegram, 2013; Klint, 2013).

A major collection edited by Musa and Dimmock (2013) published research on consumer behaviour, market segmentation and site/diver management and the environmental impacts of divers, but overall, dive firm operations have been little studied. Dimmock (2004) and Dimmock *et al.* (2013) are the major exceptions. Dimmock (2004) found that managers typically had to balance (often competing) roles, and the industry's structure and the rising demand for mass participation in diving resulted in intense price competition between firms.

Furthermore, Townsend (2008) observed that the high capital requirements for specialist equipment (diving gear, air tanks, compressors, dive boats) was highly restrictive for dive businesses. This is significant because dive operations have predominantly been small and medium-sized businesses with often more limited access to capital than larger firms. Dimmock *et al.* (2013) suggest that the key components of the dive industry can be categorized as the 'four Es' – equipment, education, experience and environment – and each service is 'mutually dependent on the health of each'.

Dive operations vary worldwide in scale and scope (Lew, 2013; Dimmock and Musa, 2015) with significant differences between large-scale operators such as those in the Caribbean or Red Sea and smaller business which may be co-located in resort areas. In other regions such as South-east Asia and the Pacific islands, whilst there are larger operators, smaller owner-operator dive businesses predominate. In some latter cases there appear to be elements of 'lifestyle entrepreneurship' (Ateljevic and Doorne, 2000) with some smaller dive tourism operators taking a 'hobbyist' approach to business rather than the more conventional profit-maximizing, cost-minimizing business operation of larger firms. However, Dimmock *et al.* (2013) argue that there is an industry trend of a continuing 'transition' from lifestyle entrepreneurs to more professionally managed operations and a 'new breed' of managers required to balance different demands such as creating a safety culture with environmental responsibility whilst managing to generate sufficient profits.

9.4 Dive Tourism in the Perhentian Islands

Both main Perhentian islands host tourism although Perhentian Kecil ('small Perhentian' island) mainly attracts low-budget international tourists and growing numbers of domestic/regional tourists (who are often non-divers). Perhentian Besar

Table 9.1. Dive operators' key business areas in the Perhentians (Dimmock, 2004: 77).

Generating business area	
Revenue type:	**Main markets:**
Diver education	Backpackers (and some domestic tourists)
Guided leisure dives	Backpackers (few domestic tourists)
Equipment rental	Backpackers, domestic tourists

('big Perhentian') on the other hand has an increasing number of mid-range hotels. The main customers are international backpackers/budget tourists, but some dive operators host domestic tourist groups often associated with dive shops from cities such as Kuala Lumpur (Table 9.1).

Dive tourism in the Perhentians comprises mainly basic dive training (PADI Open Water Diver certification), some further training (Advanced Open Water Diver certification) and guided 'fun dives' to reefs and nearby shipwrecks. The islands' underwater topography is mainly shallow reefs (typically a maximum depth of 20 m) providing easy, safe and sheltered conditions ideal for diver training. The islands have the largest number of PADI Open Water certifications in Malaysia (according to dive industry sources) and the tourist season is broadly May–September, coinciding with summer holidays in key European markets. At other times the islands have few tourists due to monsoon rains and rough seas. The accommodation and restaurant offer is geared towards budget travellers and domestic/regional tourists with simple accommodation such as wooden chalets and beach cafés (Jaafar and Maideen, 2012). Dive operations consist of small dive operators who are mainly independent from the accommodation businesses, but there is evidence of networking with some offering discounted packages to chalet owners. The islands have been predominantly settled by Malays from the peninsula with a resident population of around 1500 mainly living in the *kampung* (village) at the south of Perhentian Kecil, some distance from the main beaches. Malays originally from the mainland comprise the main business owners but there is also some Chinese-Malaysian ownership (Hamzah and Hampton, 2013).

9.5 Methodology

9.5.1 The approach

This chapter draws from a 2-year funded study of the socio-economic impacts of dive tourism in Malaysia (see also Haddock-Fraser and Hampton, 2012; Daldeniz and Hampton, 2013). A broadly qualitative approach was used, adapted from rapid rural appraisal (RRA). RRA is an established fieldwork technique common to international development projects. In our case, fieldwork mainly comprised a series of semi-structured, in-depth interviews with key island stakeholders. An RRA-type approach was the most appropriate since budget constraints meant that an extended fieldwork period of several months was unfeasible so that two short, highly focused visits could still generate sufficient data. Businesses were approached and once contact was made, other key respondents were found using 'snowball' sampling. They included dive shop staff and owners, dive professionals (instructors and divemasters – dive guides), other local tourism businesses (accommodation, catering, souvenir shops), tourists, villagers and local community leaders as well as Malaysian and international non-governmental organizations (NGOs; Table 9.2). In addition to the semi-structured

Table 9.2. Summary of respondents.

Number of respondents/job	Origin	Gender	Average age (years)
Malaysian nationals			
n=9 Dive centre owner/manager	Perhentian, Kota Bahru, Kuala Lumpur, Sabah	2 female 7 male	48
n=4 Snorkel centre	Perhentian, Terengganu	4 male	26
n=4 Dive instructors	Perhentian, Kuala Besut	4 male	33
n=1 Dive master	Taman Negara	1 male	40
n=3 Compressor operator	Perhentian, Terengganu, Kota Bahru	3 male	23
n=3 Boatman	Perhentian, Malacca	3 male	28
n=5 Accommodation owner/manager	Perhentian (guest house) Kelantan, Kuala Lumpur (chalets)	2 female 3 male	43
n=4 Shop owners [general store, travel agency, internet shop]	Perhentian, Kuala Lumpur, Kuala Besut	2 female 2 male	42
Foreign respondents			
n=1 Dive centre owner	Sweden	1 male	35
n=22 Dive instructors	UK, Canada, USA, Hong Kong, Germany, Sweden, Switzerland, Netherlands	8 female 14 male	31
n=4 Divemaster and Divemaster trainees	UK, Sweden, Denmark	3 female 1 male	23
n=2 Shop owners (jewellery)	Spain, Sweden	2 female	28
n=14 Tourists (divers and non-divers)	UK, Sweden, Peru, Germany, Spain, USA, Canada, France, Netherlands, Malaysia	5 female 9 male	29
Total respondents interviewed: n=76			

interviews, we also undertook participant observation. Four researchers (including one author) were qualified scuba divers themselves so that participant observation was possible both onshore at the dive businesses before/after dive trips, as well as on dive boats and under water. Detailed notes were taken and fed into the nightly team meetings. In addition digital mapping and photography were carried out to record tourism infrastructure.

The concept underpinning fieldwork was to 'listen to local voices' (Chambers, 1997). The research team consisted of Malaysian and British academics, and research assistants from both countries. This had two main advantages. First, several team members had prior research experience of the island, what Pagdin (1989) calls 'pre-knowledge'. Second, a team comprising both UK and Malaysian nationals allowed interviews to be undertaken either in English or *Bahasa Malaysia*. A total of n = 76 semi-structured interviews were completed in the Perhentians with average duration of 45 minutes. For this chapter n = 62 interviews were analysed as they represented tourism entrepreneurs as either owners, managers or dive instructors/divemasters. The remaining n = 14 were tourists and conservation workers not engaged in tourism or dive tourism businesses and therefore excluded.

9.5.2 Data collection and analysis

Project design enabled two fieldwork rounds ('shoulder' and peak season). Interviews were recorded in field notebooks, typed up, coded and transcripts analysed using NVivo software. The material was coded to differentiate respondents by their personal attributes (nationality, age, gender), business operations (type, size) and personal experience of working in dive tourism. This facilitated sorting using NVivo. Interviews were initially analysed by the British researchers and results then discussed with our Malaysian colleagues. Concerning reducing researcher bias, this had two aspects. First, for the fieldwork, the interview protocol was designed, and then revised by the lead academics (Malaysian and British) so that questions were structured to not 'lead' respondents to answer in certain predetermined ways and that answers could be 'triangulated' between different respondents. This gave our research team confidence in the interview material generated. Second, at the coding stage, the coding was agreed by the lead academics giving confidence that researcher bias could be minimized given that it is very difficult – perhaps almost impossible – to eliminate bias completely. One experienced researcher was allocated the task of data analysis using NVivo, but with discussion with team colleagues before and after the analysis stage.

9.6 Discussion

In the 1990s Perhentian dive businesses were owned and operated by small family units as an additional income source alongside traditional livelihoods. Often these micro-enterprises were established to meet demand from backpackers for a 'room to rent' and meals during their island stay (Hamzah, 1997). The relatively small groups of backpackers provided little incentive for business owners to strategically position themselves in competition with one another. However, by the late 2000s this had changed with a decline in seasonal entrepreneurs. The main business operations observed during fieldwork were going through stages of adapting to a highly competitive environment with high volume, low-cost markets (specifically, Open Water dive training) of not only international but also domestic tourists. Increasing tourist numbers, as well as the diversity in tourist types, presented entrepreneurs with a very different operating environment. Yet, despite increasing price competition among island businesses serving both dive and beach tourists, there was a lack of product innovation and differentiation, with most operators imitating rather than innovating. Furthermore, the government drive to support innovation in tourism and raise the opportunities for 'local communities and small entrepreneurs' in the Ninth Malaysia Plan (2006–2010) appeared to make little difference in the island.

9.6.1 Personal factors

The reasons given by the majority of respondents (not born in the Perhentians) for establishing a business in the islands was due to the relatively undeveloped tourist infrastructure and excellent diving conditions. Most were knowledgeable about neighbouring dive destinations in Malaysia and Thailand and often highly critical of the large-scale, luxury operations there, preferring the quiet, relaxed environment of the islands and the low volume, backpacker market.

The principal motivation for non-islanders was 'lifestyle' as the islands offered a way of life that typically contradicted their former city lives: 'Actually it could have been anything, any business. I love this island. I came here 3 years ago. This is the place I want to be' (male, Internet shop owner from Kuala Lumpur); and in Europe: 'It's relaxed and not all about profit. It's not all about money and stress' (male, trainee divemaster from Denmark).

For island residents, the motivation to participate in tourism was different and focused on improving their quality of life and that of their families: 'We used to have a hard living and now it is much easier compared to those days. Earlier it was hard to get a job but now I am earning by running this shop' (male, general store owner from Perhentian). For the younger generations who ran snorkelling and boat trips or worked as shop assistants or waiting staff (Table 9.2), the opportunity to become independent from the family but remain close to their home *kampung* was a strong motivator.

Aside from motivation, other personal factors were present and influencing the early stages of the entrepreneurial process. Typically, capital to start up and sustain business was from personal savings, family members or other dive- and non-dive-related businesses in the Perhentians or major cities. However, there were significant differences in the scale of business operations between local entrepreneurs and those from elsewhere. For example, those originating from Malaysian cities were typically better educated (degree level) than local people with prior experience in business ownership, and with access to capital. The high capital requirements (e.g. specialist dive equipment and boats) normally associated with dive tourism had not been observed in the Perhentians until more recently, and specifically among the *Bumiputra* from the islands and Chinese-Malaysian or Indian-Malaysian from Kuala Lumpur setting up dive shops: 'This business [dive shop] was my birthday present. My boyfriend's father owns the chalets. We ordered the equipment and then we buy a new 150hp boat' (female, dive shop owner from Sabah). The remaining non-islanders were less fortunate and the opportunity to invest time and money in establishing a dive tourism business was influenced by dive tourism's seasonal nature in the islands. For example, a second income or business supported individuals during the monsoon season providing income until the next season. This seemed to be particularly prevalent with instructors and divemasters who either returned to the city, 'I still have a job in KL [Kuala Lumpur] as technician. Off-season – set up, repair and prepare for next season' (male, instructor from KL), or gained employment with dive centres elsewhere in Malaysia or Thailand. There were fewer dive instructors and divemasters from Malaysia however, because the PADI qualification is cost prohibitive and limited to those with good English. It was explained that there are 'No PADI materials in Bahasa Malay' (male, dive instructor from the UK) thus most support staff such as boatmen, local compressor operators and waiters were local people from the region, although not the Perhentians specifically. During the monsoon season these non-islanders either returned to their *kampungs*, took up fishing locally or sought tourist-related work in Langkawi, while most from the island stayed to repair chalets or boats, to provide security for dive shops closed for the season, or to take up fishing. However, the income earned from these jobs was usually less – and in some cases half – of the earnings made from dive tourism.

Our findings show after nuanced analysis that personal factors influenced the dive tourism entrepreneurial process considerably, and determined the nature and pace of development. There was a clear distinction between those wishing to improve their quality of life and reduce precarity (the islanders), and those pursuing a 'lifestyle' opportunity that was inherently linked to respondents' origin, ethnicity and financial capital.

9.6.2 Sociological factors

The entrepreneurial process was deeply embedded in the family and social network and observed with most Malaysian respondents regardless of their business operations or personal circumstances. However, there were distinct differences in how the process evolved over time when comparing local dive tourism entrepreneurs with other Malaysians.

The initial stages of creation and innovation for local entrepreneurs were influenced by their social network and the social and cultural norms of a fishing community, not by any desire to be competitive, to profit from the business or grow the business. 'We are from this island. We have been operating for over 20 years and we started since 1990. [We] follow other landowners who start their business with small chalets. During the tourist season we operate the chalets and during the monsoon I will go back to the village . . . We are fisherman at the village' (female, accommodation owner from Perhentian).

Local, multiple business owners had strong family networks where all members (across many generations) had some experience in tourism as a worker or business owner/manager. Stages of setting up a new business were interwoven with maintaining the family unit and maximizing existing equipment and manpower, rather than expanding their business' capacity to compete with larger operators: 'This is my family business [restaurant] and I am helping them [siblings] to run the [other] businesses . . . we are focus on the restaurant business and at the same time we also provide the water taxi services. We start to do snorkelling trip using our existing boat' (male, boatman from Perhentian). This suggests that dive tourism in the Perhentians remains low-cost for local business owners who continue to serve a predominantly international, budget market. Furthermore, the motivation to improve 'quality of life' and reduce precarity with local business owners extends beyond the self to the whole family unit. Thus, the choice to diversify rather than grow operations is a strategy that ensures all family members can improve their quality of life.

Although the family unit was equally important to other Malaysian respondents originating from cities, they also displayed characteristics more associated with the 'western entrepreneur' to innovate and to profit from that innovation: 'We opened this year. It's our first year. It's a family business. I'm looking into tourism line. Looks like a lot of profit. We want to fulfil the need. Our customers request diving and snorkelling' (male, dive shop manager, Kota Bahru). In this particular case the respondent's aunt and uncle were qualified dive instructors and the business's expansion provided job security for remaining family members. There were few respondents in this category who had followed friends but rather experienced the islands as a tourist or gained knowledge through dive tourism networks in Kuala Lumpur, for example. This was also noted with dive professionals from other Asian countries: 'Come here with working visa. I used to come with family, for holiday. Many times. They know this place' (male, dive instructor, Hong Kong).

A similar route was taken by international respondents, and mainly Europeans; however, without the family unit close by there was an observable camaraderie among them that seemed to fill this void. The small, close network of predominantly dive professionals were very mobile (moving to other locations during the monsoon season) but also highly connected (via social media, previous work connections), therefore it was not uncommon to 'follow a friend' to another dive destination to freelance. Although the decision to move during the monsoon season was financially motivated, the overriding factor was to maintain a lifestyle of diving in great conditions 'among friends' (the diving community).

The prevalence for local businesses to follow their peers and imitate business models was similarly observed with younger generations from the region who engaged in low-skilled operations (compressor operator, boat operator, shop assistant) and typically 'followed a friend' (male, local, compressor operator) to learn the trade. These individuals had limited business experience and fewer qualifications than Malaysians from large cities, and they relied upon their friends and/or family to learn skills in different aspects of the dive tourism business. This was not to gain employment with one of the established island businesses however, but to become freelance operators in their own right. This approach was taken by most young regional entrepreneurs as it gave them autonomy to balance work commitments in the Perhentians and family obligations in neighbouring islands or regions.

The entrepreneurial process in all cases was strongly influenced by a close bond between family members or the diving community, and there was an inherent desire to maintain this equally with the business or businesses. Thus, despite a transitioning tourism economy on the island and increasing competition, the existing social networks and cultural norms were resilient and able to resist change. What is more interesting is how different groups of owners resisted the change. Those motivated by improving 'quality of life' were more creative and versatile with their resources and manpower, preferring to diversify services for outsourcing to dive shops owned by Malaysian nationals. They, in turn, were motivated by 'lifestyle' factors, preferring to run well-equipped but small, profitable operations to complement the family network and ambitions. Although financial capital did not particularly constrain this group, there was no explicit drive to change the business to the high-cost model observed in larger dive destinations elsewhere.

9.6.3 Environmental factors

There was a combination of internal and external environmental factors that influenced how the entrepreneurial process evolved over time. First, local owners cited difficulty in accessing bank loans for business start-ups, or formal business support: 'there is no encouragement from government' (male, dive shop owner from the Perhentians). Thus, many entrepreneurs had invested personal or family funds into the business and adapted existing equipment (such as fishing boats), skills and knowledge for dive tourism purposes. The trend to use personal savings was also observed among Malaysian nationals who had extended business networks (often in major cities) and past business experience. Unlike local owners, financial capital was not an observable challenge for Malaysian owners who could invest in new equipment for their businesses, and who could potentially access bank loans given their family and business ties in cities. However, there was no desire to expand business beyond the means of the family and the lifestyle from running a small business.

The lack of formal support and the informal business environment in the Perhentians helps explain why expertise, equipment and manpower was often shared between dive tourism businesses and why competition was viewed as 'friendly', in spite of price cutting by some: 'Everyone gets on well with each other. It's a healthy camaraderie. There is a little bit of a price difference, but it's not significant' (male, dive instructor from the UK).

A second external environmental factor was the impact of rising fuel prices on daily business operations, which presented significant challenges for businesses to maintain consistency with increasing operational costs. Dive shops for example, promoted particular dives that required less fuel for dive boats – 'we try to have fun dives nearby, make it a higher priority' (male, dive instructor from Canada), and were able to maintain business as usual. Alternatively, accommodation owners were unable to maintain the same standard of service without increasing costs, and so rationed electricity usage to coordinate with their guests' movements – air conditioning and Internet access (if provided) was restricted to evenings and mornings for example, when demand was greatest. Those operating both accommodation and a dive shop divided the supply to meet demand: 'the dive shop have it [electricity] during the daytime because they need to watch videos regarding diving and to compress [air for] the tank' (female, dive shop owner from Perhentian). At night, the generator was switched to supply the accommodation and restaurant businesses. Despite the challenges, island businesses were sufficiently resilient to maintain business operations and customer satisfaction.

The development of a large new jetty was the third factor that influenced the entrepreneurial process because it triggered action to either diversify or re-evaluate their business direction. Unlike the external factors (accessing bank loans or managing rising fuel costs), the disruption from the jetty was significant, interrupting their business operations in several ways. First, the completed structure obstructs views of the sunset – a key selling point for many accommodation and restaurant businesses – which cannot be resolved through any business strategy given the jetty's permanence. Second, water taxi operators lost business and some had to abandon the enterprise to seek alternative business ventures with their boats: snorkelling tours or outsourcing for dive centres were the preferred options. For other local and regional boat operators however, business ownership was abandoned altogether by selling their boat, returning to fishing, waiting on tables in restaurants or leaving the island for better opportunities. Third, the multiple family businesses, in contrast, had sufficient support to quickly utilize the boats within existing dive operations or add snorkelling tours as a new service. However, the decline in water taxi and freelance boat operators was a challenge for smaller businesses who relied on outsourcing services. The physically dominant, oversized and permanent presence of the jetty was the most influential factor disrupting the later stages of the entrepreneurial process, or in some cases terminating the process. Yet despite this, and the potential business opportunities from the new jetty, business owners continued to manage this change by reverting to small, 'tried and tested' business operations.

Employment was the final environmental factor affecting the business operations of Malaysian nationals, who typically offered a high quality product but employed locally where possible. However according to one male accommodation owner from Kuala Lumpur, local people 'prefer to freelance. They don't want to work in fixed job. They want their own business, own shops. When I asked them, they say "I want [to] be on my own."' The smaller pool of local employees over the larger group of local tourism entrepreneurs was significant enough for businesses to adapt to changing trends by choosing to outsource services, rather than own certain equipment (for example, boats) and employ workers (such as compressor operator) – a strategy that was challenged with the jetty development. Furthermore, local employees were also considered by some to be 'lazy' or 'aloof', however this was a misguided observation as one local snorkel shop owner explained: 'before this, I work as a waiter and getting a good pay about RM1000 [US$270]. But, I had problem to wake up early because they start to operate from 7am to 11pm. Then I decide to work by my own, buy a boat and snorkelling equipment and built the small stall there' (male, from Terengganu). In this and other cases the unsociable and seasonal working hours were obvious challenges for local people with a different way of life. This was a key reason for starting a dive or snorkel business but only after experience was acquired gradually as an employee with one or multiple businesses: 'Since 1998 . . . I started as [a] compressor guy; then became a boatman and took my diving courses and now I'm an instructor here. I start my career with [name of dive operator] and got training from them and did my Open Water level with [another dive operator]' (male, local dive instructor from Perhentian). Although employment was a critical factor for the larger businesses, it was less of an issue for smaller, local owners and family businesses who could draw on the community and family network for assistance, thus continuing operations 'as normal'.

9.7 Conclusions

This chapter explored *how* the tourism entrepreneurial process transforms or not, by examining the critical factors (personal, sociological and environmental) that influence such change. Dive tourism was selected because it represents a growing and important market for coastal and island regions in LDCs; once serving a relatively small group of specialist tourist such as backpackers, it now caters for mass tourism.

The key findings from our research in Malaysia broadly showed dive tourism businesses in Perhentian remained small, despite growing competition and price increases. Furthermore, there was a tendency towards imitating rather than innovating products and services (also noted by Koh and Hatten, 2002) even when opportunities arose. These initial observations partly concur with prior research that found tourism entrepreneurs were more inclined to reject growth and profit models over models that fulfilled a lifestyle or non-financial need (Jaafar *et al.*, 2011; DeBerry-Spence and Elliot, 2012). However, this was only evident among individuals with experience of modern capitalism as a former resident and/or business owner in a major city or developed country. The main motives for local and regional entrepreneurs with limited experience of the capitalist model correlated strongly with quality of life (not lifestyle) goals, which were inherently associated with the family, the community network and the island's marine environment. Thus, the key factors influencing the entrepreneurial process were, initially, embedded in the social and cultural norms of an island fishing community, rather than originating from transitioning economic and business norms of dive tourism operations in the Perhentians.

The process undertaken by many respondents, during the initial stages of the entrepreneurial process, was guided by friends or family who encouraged, trained or financed those first steps. There was no evidence that government intervention influenced the decision to start a tourism business, nor was there evidence that bank loans were available to fund operations, subsequently what is commonly associated with the dive tourism sector – equipment-intensive (Dimmock *et al.*, 2013) and high capital investment (Townsend, 2008) – was not evident in this case. There were some differences between local and non-local entrepreneurs in how the early stages were managed, and largely defined by their personal circumstances and financial capacity. Non-local Malaysian entrepreneurs had more start-up capital and were able to equip their businesses with a larger stock of specialist dive equipment for rent than smaller, local operators. This increase in capacity generated increased income from more dive tourists and from renting equipment to smaller businesses. Despite the potential business opportunity to supply and sell dive equipment to dive tourists and local dive businesses – particularly with growing visitor numbers – there was no observable inclination to follow through. This contradicts the findings from Dimmock *et al.* (2013), who identify this activity as a key component of the dive tourism businesses.

For local entrepreneurs from the islands, they either had training in business from their family or from working in other businesses. In addition there was evidence of some 'laddering' experience in some, or all aspects of the dive tourism business and this resulted in the start-up of their own business. Our findings concur with Lee *et al.*'s (2011) research of Malaysian entrepreneurs as having a strong sense of autonomy, self-reliance and resilience, but there was no observable risk-taking.

Concerning sociological factors, these appeared more complex and were interwoven with internal and external environmental factors, which influenced the process significantly in the mid- and latter stages of the entrepreneurial process (triggering event, implementation and growth).

Dive tourism has been propelled by local tourism entrepreneurs and small businesses and our fine-grained analysis demonstrates how the tourism entrepreneurial process in an island destination has been clearly influenced by personal, sociological and environment factors, some of which arise from Malaysia's historical specificity. Further comparative work on the tourism entrepreneurial process would be useful to test this notion in other regions such as the Caribbean, Indian and Pacific Oceans, and also cross-comparisons could be investigated with SIDS and other island economies. Specific aspects for further research could include access to capital and the role of the state in relation to tourism entrepreneurs in LDCs. Finally, our research has practical implications for LDC policy-makers and international NGOs and development organizations since tourism remains a key driver within strategic planning for economic growth, employment creation, poverty alleviation and the transformation of local livelihoods.

Acknowledgements

This chapter is an output from the PMI2 Project funded by the UK Department of Business, Innovation and Skills (BIS) for the benefit of the Malaysian Higher Education Sector and the UK Higher Education Sector. Any views expressed are not necessarily those of BIS, nor British Council. We would like to thank our project colleagues Janet Haddock-Fraser, Amran Hamzah and Ghazali Musa. In addition we thank Lngesh, Bilge and Jorn for their assistance with fieldwork, as well as the interview respondents who generously gave their time. We would also like to thank Julian Clifton, Kay Dimmock, Mark Gilman and Paul Philips for their helpful comments on earlier versions of the chapter. Finally we are grateful to Perseverence Jeyacheya for turning sketch maps into figures. The usual disclaimers apply.

References

Ahmad, S.Z., Jabeen, F. and Khan, M. (2014) Entrepreneurs choice in business venture: motivations for choosing home-stay accommodation businesses in Peninsular Malaysia. *International Journal of Hospitality Management* 36: 31–40.

Ariff, M. and Abubakar, S.Y. (2005) Strengthening entrepreneurship in Malaysia. Available at: http://www.mansfieldfdn.org/backup/programs/program_pdfs/ent_malaysia.pdf (accessed 30 June 2016).

Ateljevic, J. and Doorne, S. (2000) Staying within the fence: lifestyle entrepreneurship in tourism. *Journal of Sustainable Tourism* 8(5), 378–392.

Basiron, M.N. (1994) Marine tourism in Malaysia and its potential. Paper read at National Conference on the Contribution of the Maritime Sector to the National Economy, Kuala Lumpur, 30 November.

Bjerke, M.V. (2000) A Typified, Culture-based, Interpretation of Management of SMEs in Southeast Asia. *Asia Pacific Journal of Management* 17, 103–132.

Bygrave, W.D. (1993) Theory building in entrepreneurial paradigm. *Journal of Business Venturing* 23, 255–280.

Bygrave, W.D. (2004) The entrepreneurial process. In: Bygrave, W.D. and Zacharakis, A. (eds) *The Portable MBA in Entrepreneurship*. John Wiley & Sons, Hoboken, New Jersey, pp. 1–27.

Canty, S. (2007) Positive and negative impacts of dive tourism: the case study of Utila, Honduras. Unpublished MSc thesis. Lund University, Lund, Sweden.

Chambers, R. (1997) *Whose Reality Counts? Putting the First Last.* Intermediate Technology Publications, London.

Clifton, J. (2004) Evaluating contrasting approaches to marine ecotourism: 'dive tourism' and 'Research tourism' in the Wakatobi Marine National Park, Indonesia. In: Boissevain, J. and Selwyn, T. (eds) *Contesting the Foreshore. Tourism, Society, and Politics on the Coast.* Amsterdam University Press, Amsterdam, pp. 151–168.

Commonwealth Secretariat (2002) *Jamaica Master Plan for Sustainable Tourism Development.* Commonwealth Secretariat, London.

Daldeniz, B. and Hampton, M.P. (2013) Dive tourism and local communities: active participation or passive impacts? Case studies from Malaysia. *International Journal of Tourism Research* 15(5), 507–520.

Dana, L.-P., Gurau, C. and Lasch, F. (2014) Entrepreneurship, tourism and regional development: a tale of two villages. *Entrepreneurship and Regional Development* 26(3–4), 357–374.

Davis, T. (2005) *Local and Semi-Local Economic Impacts of Dive Tourism in Bunaken National Park, North Sulawesi, Indonesia.* WWF-Indonesia, Jakarta.

Davison, B. (2007) How many divers are there? And why you should care. *Undercurrent Magazine* 22(5) May issue. Available at: http://www.undercurrent.org/UCnow/dive_magazine/2007/HowManyDivers200705.html (accessed 9 June 2014).

DeBerry-Spence, B. and Elliot, E.A. (2012) African microentrepreneurship: the reality of everyday challengers. *Journal of Business Research* 65, 1665–1673.

Dimmock, K. (2004) Managing recreational scuba experiences: exploring business challenges for New South Wales dive tourism managers. *Tourism Review International* 7(2), 67–80.

Dimmock, K. and Musa, G. (2015) Scuba diving tourism system: a framework for collaborative management and policy. *Marine Policy* 54, 52–58.

Dimmock, K., Cummins, T. and Musa, G. (2013) The business of scuba diving. In Musa, G. and Dimmock, K. (eds) *Scuba Diving Tourism.* Routledge, London, pp. 161–173.

Edney, J. (2012) Diver characteristics, motivations and attitudes: Chuuk Lagoon. *Tourism in Marine Environments* 8, 7–18.

Garrod, B. (2008) Market segments and tourist typologies for diving tourism. In: Garrod, B. and Gossling, S. (eds) *New Frontiers in Marine Tourism.* Advances in Tourism Research Series. Elsevier, London, pp. 139–152.

Gartner, W.B. (1990) What are we talking about when we talk about entrepreneurship? *Journal of Business Venturing* 5(1), 15–28.

Haddock-Fraser, J. and Hampton, M.P. (2012) Multi-stakeholder values on the sustainability of dive tourism: case studies of Sipadan and Perhentian islands, Malaysia. *Tourism Analysis* 17(1), 27–41.

Hampton, M.P. and Jeyacheya, J. (2013) *Tourism and Inclusive Growth in Small Island Developing States.* Commonwealth Secretariat, London.

Hamzah, A. (1997) The evolution of small-scale tourism in Malaysia: problems, opportunities, and implications for sustainability. In: Stabler, M.J. (ed.) *Tourism and Sustainability: Principles and Practice.* CAB International, Wallingford, UK, pp. 199–217.

Hamzah, A. and Hampton, M.P. (2013) Resilience and non-linear change in island tourism. *Tourism Geographies* 15(1), 43–67.

Hillmer-Pegram, K. (2013) Understanding the resilience of dive tourism to complex change. *Tourism Geographies*, DOI: 10.1080/14616688.2013.851268

Hollick, M. and Braun, P. (2005) Lifestyle entrepreneurship: the unusual nature of the tourism entrepreneur. *Proceedings of the second annual AGSE international entrepreneurship research exchange.* Swinburne Press, Melbourne.

Jaafar, M. and Maideen, S.A. (2012) Ecotourism-related products and activities, and the economic sustainability of small and medium island chalets. *Tourism Management* 33, 683–691.

Jaafar, M., Abdul-Aziz, A.R., Maideen, S.A. and Mohd, S.Z. (2011) Entrepreneurship in the tourism industry: issues in developing countries. *International Journal of Hospitality Management* 30, 827–835.

Klimmek, H. (2013) An examination of the barriers to local community participation in the dive tourism industry in Flores, Indonesia. Unpublished MA. King's College, London.

Klint, L.M. (2013) Buoyancy – Bifo and Afta: A climate change vulnerability/resilience framework for tourism – the case study of Vanuatu dive tourism. Unpublished PhD. Victoria University, Melbourne.

Kodithuwakku, S.S. and Rosa, P. (2002) The entrepreneurial process and economic success in a constrained environment. *Journal of Business Venturing* 17, 431–465.

Koh, K.Y. and Hatten, T.S. (2002) The tourism entrepreneur. *International Journal of Hospitality and Tourism Administration* 3(1), 21–48.

Lee, S.M., Lim, S.B. and Pathak, R.D. (2011) Culture and entrepreneurial orientation: a multi-country study. *International Entrepreneurship and Management Journal* 7(1), 1–15.

Lew, A. (2013) A World Geography of recreational scuba diving. In: Musa, G. and Dimmock, K. (eds) *Scuba Diving Tourism.* Routledge, London.

McKercher, B. (1999) A chaos approach to tourism. *Tourism Management* 20, 425–434.

Ministry of Tourism, Arts and Culture (2012) Fourth Tourism Master Plan 2013-17. Volume 2: Background and Analysis. Draft 1. Ministry of Tourism, Art and Culture, Maldives. Available at: http://www.tourism.gov.mv/downloads/news/2012/4TMP_DRAFT_Vol_2_BACKGROUND_ANALYSIS.pdf (accessed 9 June 2014).

Mograbi, J. and Rogerson, C.M. (2007) Maximising the local impacts of dive tourism: Sodwana Bay, South Africa. *Urban Forum* 18(2), 85–104.

Morris, M.H. and Lewis, P.S. (1991) Entrepreneurship as a significant factor in societal quality of life. *Journal of Business Research* 23, 21–26.

Morrison, A. (2006) A contextualisation of entrepreneurship. *International Journal of Entrepreneurial Behaviour and Research* 12(4), 192–209.

Morrison, A., Rimmington, M. and Williams, C. (1999) *Entrepreneurship in the Hospitality, Tourism and Leisure Industries.* Butterworth-Heinemann, Oxford, UK.

Musa, G. (2002) Sipadan: a SCUBA diving paradise: an analysis of tourism impact, divers satisfaction and tourism management. *Tourism Geographies* 4(2), 195–209.

Musa, G. (2003) Sipadan: an over-exploited scuba-diving paradise? An analysis of tourist impact, diver satisfaction and management priorities. In: Garrod, B. and Wilson, J. (eds) *Marine Ecotourism. Issues and Experiences.* Channel View, Clevedon, UK, pp. 122–137.

Musa, G. and Dimmock, K. (eds) (2013) *Scuba Diving Tourism.* Routledge, London.

Ong, T.F. and Musa, G. (2011) SCUBA divers' underwater responsible behaviour: can environmental concern and divers' attitude make a difference? *Current Issues in Tourism* 15(4), 329–351.

PADI (2014) *Worldwide Corporate Statistics 2014. Data for 2008-2013.* Professional Association of Diver Instructors, Rancho Santa Margarita, California. Available at: http://www.padi.com/scuba-diving/about-padi/statistics (accessed 10 September 2014).

Pagdin, C. (1989) Assessing tourism impacts in the third world: a Nepal case study. *Progress in Planning* 44(3), 185–226.

Ryan, T., Mottiar, Z. and Quinn, B. (2012) The dynamic role of entrepreneurs in destination development. *Tourism Planning and Development* 9(2), 119–131.

Scheyvens, R. (2006) Sun, sand, and beach *fale*: benefiting from backpackers - the Samoan way. *Tourism Recreation Research* 31(3), 75–86.

Timmons, J.A. (1999) *New Venture Creation: Entrepreneurship for 21st Century,* 5th edn. Irwin-McGraw-Hill, Homewood, Illinois.

Tourism Malaysia (2009) Tourism Minister: be creative and innovative in attracting tourists. Press Release, 13 July. Tourism Malaysia, Putrajaya. Previously available at: http://corporate.tourism.gov.my/mediacentre.asp?page=news_desk&news_id=357&subpage=archive (accessed 6 March 2014).

Tourism Queensland (2003) Dive tourism. Fact Sheet. Tourism Queensland, Brisbane. Previously available at: http://www.tq.com.au/fms/tq_corporate/research/fact_sheets/dive_tourism.pdf (accessed 9 June 2014).
Townsend, C. (2008) Dive Tourism, Sustainable Tourism and Social Responsibility: A Growing Agenda. In: Garrod, B. and Gössling, S. (eds) *New Frontiers in Marine Tourism: diving experiences, sustainability, management*. Elsevier, London, pp. 189–200.
UN WTO (2015) *Tourism Highlights*. UN World Tourism Organization, Madrid.
Worachananant, S., Carter, R.W., Hockings, M. and Reopanichkul, P. (2008) Managing the impacts of SCUBA divers on Thailand's coral reefs. *Journal of Sustainable Tourism* 16(6), 645–663.
WTTC (2015) *Economic Impact 2015 Malaysia*. World Travel and Tourism Council, London.

10 Stakeholders' Perceived Impact of Wind Farms at a Tourism Destination: A Case Study of Jeju Island, South Korea

SEUNGWOO JOHN LEE*

Sangmyung University, South Korea

10.1 Introduction

In today's society, energy consumption around the world is an increasingly concerning issue, using up vast amounts of the Earth's natural resources. As we progress further into the 21st century, many nations around the world are realizing the importance of alternative sources of energy in response to increasing energy consumption and the possible negative consequences resulting from traditional energy sources such as coal and gas. Renewable energies do not simply rely on coal and gas, or even nuclear power, rather the energy sources are from natural resources, such as geothermal, sun and wind. Therefore, many governments have undertaken in recent years building renewable power plants as solar arrays, geothermal power plants and wind turbines (Dincer, 2000).

In particular, alternative energy development is deemed to be necessary in many small islands due to the heavy dependence on oil importation or locally generated weak electricity (Duic and Carvalho, 2004). Tabatchnaia-Tamirisa *et al.* (1997), for example, claim that Hawaiian energy consumption continues to rise and recommend that alternative energy resources should be implemented to serve growing travel markets.

Concerning the large scale development, multiple stakeholders are involved to facilitate the process. In that regard, understanding multiple stakeholders is an integral part of the decision-making process to the typical groups of residents and government officials (Merrilees *et al.*, 2012). In addition to having stakeholders involved, congruity between them can be crucial for such a large development because the development can create conflicts when things are not in the same direction and one party may feel pressure to change his/her mind over the other side (Osgood

*E-mail: slee@smu.ac.kr

and Tannenbaum, 1955). In the meantime, such a large scale development can bring disparate stakeholders to communicate and interact with each other when plans are put together collaboratively (Wordwiz, 2013).

Even though there are studies that examined various perceptions of wind farms, little research exists that has examined perceptions from multi-stakeholders' perspectives simultaneously. There are even fewer studies when wind farm development takes place at a tourism destination and when it comes to the relationship between tourism and wind farms.

The purpose of this study is to explore the opinion of multi-stakeholders regarding tourism preferences and wind farm development using stakeholder theory and congruence theory. To address the purpose of this study, this study includes two stakeholder groups, residents and government officials. The following questions are proposed:

1. How and in what way do stakeholders express their perceptions toward tourism preferences and wind farm development?
2. What factors, such as economic, environmental and cultural, etc., influence stakeholders from being supportive or in opposition to wind farm development?

10.1.1 Tourism and stakeholder theory

Stakeholder theory consists of an idea of how business is operated. The theory emphasizes the importance of the relationship that firms have with stakeholders. An integral part of generating a successful business is the realization that the stakeholders are people, not functions (Jensen, 2001). With a higher number of people, there is greater likelihood of discord for the overall goal. So creating value can be achieved by involving various stakeholders' voices. In addition, though comprised of separate parts, Phillips (2004) emphasizes the need to look at the groups as one cohesive unit interrelating. By doing so, they can create a successful business together, which otherwise would be impossible (p. 3).

In particular, the combination of stakeholders may be more crucial in the tourism industry than other industries because it comprises a sophisticated network of stakeholders that are interested in development, management and promotion of tourism at a destination (d'Angella and Go, 2009). In regard to stakeholder's involvement in the tourism industry, Timur (2010) adds that engaging stakeholders brings many potential benefits; in particular, significant engagement helps promote development, improvement and success of an attraction or destination by increasing efficiency as well as decreasing social, political, economic and environmental concerns. Timur (2010, p. 1) states that stakeholders' engagement is a key part of achieving ultimate business success and can 'provide high-quality visitor experiences that can maximize the benefits to destination stakeholders'.

Despite the fact that the involvement of stakeholders seems to be mutually beneficial, reality does not always live up to this ideal system because each stakeholder wants his/her voice to be heard more. It is evidenced that residents' voices often tend to remain unheard in regard to tourism development because tourism developers and government often make a decision and notify the decision to residents (Kaplanidou and Vogt, 2003). As a result, conflicts between stakeholders only grow. An example that represents conflicts between stakeholders pertaining to tourism development is seen in the study of Davis and Morais (2004) with regard to unsustainable tourism development in Williams, Arizona. They argue that tourism development is tangled with several attributes that lead to conflict. In the study, for example, local government has to abide by city ordinances, which take much time to review and apply to proposed development, while the tourism enterprises want to operate the business immediately. While the relationship between the two stakeholders has created conflicts, local residents remain to be left out in the decision-making process. This demonstrates that tourism development operated by any one-sided stakeholder creates conflict and dissention.

10.1.2 Tourism and congruence theory

In the similar vein, the importance of the relationship between stakeholders is evidenced in congruence theory. The congruence model is the principle that an organization's performance is derived from multi-attributes associated with an organization (Eckstein, 1997). This idea suggests that the higher the degree of congruence amongst these attributes, the greater the chances of increased performance. For example, if an attribute in a firm stands out, but the other attribute is not a good fit for the way the firm directs, their congruence will not be achieved. Conversely, contributions of this theory lie in the fact that the need of involving public and private sectors gathers consensus to reach the same goals (Jamal and Getz, 1995; Bramwell and Lane, 2000). In particular, a sustainable tourism development requires collaborative and mutual building relationships (Healey, 1997). In order to facilitate the consensus strategy, Healey (1997, p. 244) suggests that openness and trust can lead to consensus-building relationships by identifying 'who is involved, who should be involved and who may be left out'. Healey adds that effective collaboration from disparate groups helps find problems and provide strategies and directions.

An integral part of generating a successful development depends on how congruent the stakeholders are. As noted previously, greater likelihood of unbalance increases with a higher number of units. So what matters is, though comprised of separate units, how to involve various stakeholders' voices into one cohesive unit (Phillips, 2004). Although combining various opinions into one voice can be difficult, Eckstein (1997) explains that similarity between different associations represents congruence. Chang and Huang (2011) propose the typology that leads to clarification of the role of each stakeholder for urban reclamation in Singapore. By establishing the clear role, each stakeholder understands their responsibility and the ultimate goal, which in turn increases the likelihood of congruence. This study suggests that congruence does not always force stakeholders to be united; rather understanding and respecting the role of each stakeholder can be a norm and may entail an ideal state of affairs (Merrilees *et al.*, 2012).

From the review of previous studies, it is sensed that identifying and understanding who are the stakeholders and what are their roles and interests can be important in engaging all entities in the tourism development process.

10.1.3 Wind farms

In response to growing energy consumption, many organizations turn their interest to alternative energy development. The tourism industry is not an exception because the tourism industry, for example, demands great amounts of energy at various functions such as accommodation, catering and the management of tourist attractions (Becken *et al.*, 2003). Thus, of particular interest in the tourism industry, renewable energy development receives a great deal of attention from politicians and decision makers because renewable alternative energy sources have a minimal impact not only on the environment but also on other aspects such as the quality of life and health (Cavallaro and Ciraolo, 2005).

In a similar vein of thought, this increasing growth is attributed to multifaceted factors, including economic, environmental and social impact at a destination. In a favourable aspect of wind farm development, previous research (Dalton *et al.*, 2008) found that wind farms could be a medium to diversify tourism attractions that benefit stakeholders at a destination. Local residents may benefit from wind farms. Robb (2003) found that residents are in favour of bringing wind farms within their property because they are given to reserve power generated from wind farms and could sell unused electricity, resulting in a new source of household income. Tourism generates a significant amount of revenue to local, regional and state government. In a way, government pays close attention to development that may bring tax dollars into the office (Lee *et al.*, 2013). Electricity generated from wind farms stays in the local community as opposed to leaking out to where the oil resources are imported. Cavallaro and Ciraolo (2005) demonstrate that wind farms on an island would bring considerable economic benefits to the local tourism industry.

On top of the economic benefits, positive environmental impacts are also seen. Diederichs *et al.* (2008) state that wind farms in the North Sea facilitate expansion of harbour porpoises after offshore wind farm establishment. Several scholars (Cavallaro and Ciraolo, 2005; Kempton *et al.*, 2005; Warren *et al.*, 2005) argue that wind power plants cause reduction of CO_2 emissions and replacement of an oil-burning electric plant, which in turn protects ecosystems.

Although positive impacts of wind farms are seen, it is important to note that there is evidence that stakeholders show discrepant voices. The foremost public opposition to wind farms is related to noise and visual impacts (Wolsink, 2000; Devine-Wright, 2005). Noise problems rise with regard to rotating turbine blades, which may have dangerous impacts on wildlife such as birds and animal habitat while visual negative impacts refer to an array of turbines in near a residential area. Regardless of the negative impacts, smaller wind farms are positively perceived in comparison with larger scale wind farms (Devine-Wright, 2005) in several countries including Denmark (AIM, 1993) and the UK (Lee *et al.*, 1989).

10.1.4 Characteristics of study area

Jeju is the largest island in Korea. The island presents a great deal of diversity of natural environments as a result of a combination of cultural, natural and geographical attributes. The value of island is evidenced by UNESCO designations in all three natural sciences, including Biosphere Reserve in 2002, World Natural Heritage in 2007 and Global Geopark in 2010. These features are rarely found in Asia and very few worldwide.

The island is known for its strong winds. According to Jeju province, the average wind speed as of 2003 was 3.1 m/sec and the highest speed was 6.9 m/sec around the coastline, while the wind speed on the high mountain area peaks was 9.7 m/sec. In addition, Jeju is located in the path of the summer and winter typhoons. Strong winds are often seen on the north-west coast where the winter monsoon has great influence, while the south-east coast has relatively weak winds less than 3.0 m/sec, which results in the major influx of tourists.

10.1.5 Method

Two stakeholders in the study area were selected for this study: residents and government officials. A random sampling for residents and purposive sampling for government officials were utilized; the former group of people was randomly contacted in front of the local government office and the latter group contacted through the human resources department. In the case of the second group, the author contacted the human resources manager to obtain assistance via e-mail and phone. The human resources manager agreed to distribute the call before the research team visited the office. Two research assistants along with the author conducted the survey. The author explained the study purpose and a self-administered survey was distributed to both residents and government officials. In order to avoid double participation, residents were asked not to participate in the survey more than once.

A total of 1501 households and 123 government officials were in the survey area, of which 467 residents and 106 government officials were selected for the survey. The response rate of government officials was 86%, while the response rate of resident groups was 31%.

10.1.6 Survey measurement

The questionnaire for this study was drawn from previous research (Lee *et al.*, 2013) and some variables were modified through interviews with several academic experts.

The survey consisted of four sections. The first section asked study participants to check their level of agreement toward wind farm development variables. These measurement variables focused on cultural, local, social and environmental impacts. The second section was comprised of seven tourism preference variables. The third section asked both stakeholders whether they originally supported and still support wind farm development. Some wind farms have been built and some are under construction. Thus at the time study, it is worthwhile to look whether the participants' level of agreement has changed. Questions in the three sections were measured using a 5-point Likert-type scale ranging from *strongly disagree* (1) to *strongly agree* (5). The final section contained socio-demographic questions such as age, education, gender and income.

10.2 Findings

The descriptive analysis in regard to study participants' socio-demographics revealed the following results. In terms of gender, residents comprise 45.5% female (n = 207) and 54.5% male (n = 248) while government officials consist of 34.3% female (n = 36) and 65.7% male (n = 69). The average age of study participants is 45 years, ranging from 19 to 82 years old. Marital status of all study participants is about 20% unmarried and the rest married. More than a half of study participants went to college or higher academic institutions (52.8%, n = 296). Average of length of residence reveals 30 years in their community and average household income is US$25,847.

10.2.1 Cluster analysis of tourism preferences

A cluster analysis using hierarchical and K-means clustering was employed over the seven tourism preference statements and three clusters were generated. Table 10.1 depicts the three clusters named as follows: tourism supporters (n = 244, 43%), tourism neutrals (n = 281, 49.6%) and tourism opponents (n = 42, 7.4%). The three clusters are similar to the categories used by Lee *et al.* (2013), who studied non-tourism development at a tourism destination. The mean value of the three clusters shows significant differences. Obviously, tourism supporters are highly interested in any type of tourism preferences; tourism neutrals are neutral yet somewhat positive of tourism preferences, while tourism opponents are not interested in any type of tourism preferences at all.

10.2.2 Socio-demographic variables and clusters

Socio-demographic variables are segmented to identify their characteristics against three clusters (Table 10.2). For the analysis of clusters on socio-demographic variables, chi-square analysis and analysis of variance (ANOVA) with Tukey post-hoc test were employed. As expected, mixed results are found among socio-demographic variables.

Table 10.1. Cluster analysis of tourism preferences.[a]

Tourism preferences	Means	Supporters n = 244 (43.0%)	Neutrals n = 281 (49.6%)	Opponents n = 42 (7.4%)
4. I would like to see more diverse tourists to visit our region.	4.07	4.66	3.00	1.50
6. I would like to see more international tourists to visit our region.	4.11	4.75	3.26	1.50
1. I would like to see more families visit my region for tourism.	4.02	4.55	3.32	1.00
3. I would like to see more packaged tourists visit my region.	4.18	4.64	3.32	1.50
2. I would like to see more free independent travellers/ backpackers visit my region.	3.95	4.50	3.42	1.50
5. I would like to see more tourism development in my region.	4.18	4.45	3.37	1.00
7. I would like to see more local residents involved in tourism in my region.	3.89	4.57	3.11	1.00

[a]1 being strongly disagree and 5 being strongly agree

Significant variables include stakeholders, gender, education, tourism experience, marital status, age and length of residence while household income appears to be insignificant against the three clusters.

Two stakeholders show somewhat different opinions regarding the three clusters. More than a half of residents (54.9%, n = 253) are *neutral* of tourism preferences while more than half of government officials (64.2%, n = 68) are *supportive* of tourism preferences.

In terms of gender, female groups are more neutral than supportive (37.3%) of tourism preferences (54.8%) while male groups are almost equally distributed to both 'supporters' (47.6%) and 'neutrals' (45.1%).

With respect to education, the results show that the level of education plays a significant role in determining the degree of interest in tourism preferences. As participants have a higher degree of education, except for the graduate education, they are more supportive of tourism preferences than the less educated. In particular, the group of elementary education is more 'tourism opponents' (7.9%) than 'tourism

Table 10.2. Socio-demographic variables and clusters.

	Supporters (Cluster 1)	Neutrals (Cluster 2)	Opponents (Cluster 3)
Stakeholders $p = 0.000$	Res: 38.2%, n = 176 Gov: 64.2%, n = 68	Res: 54.9%, n = 253 Gov: 26.4%, n = 28	Res: 6.9%, n = 32 Gov: 9.4%, n = 10
Gender $p = 0.050$	Female: 37.3%, n = 90 Male: 47.6%, n = 150	Female: 54.8%, n = 132 Male: 45.1%, n = 142	Female: 7.9%, n = 19 Male: 7.3%, n = 23
Education $p = 0.000$	Ele: 2.6%, n = 1 Mid: 21.8%, n = 12 High: 43.5%, n = 73 College: 52.1%, n = 151 Grad: 33.3%, n = 2	Ele: 89.5%, n = 34 Mid: 74.5%, n = 41 High: 45.8%, n = 77 College: 41.7%, n = 121 Grad: 66.7%, n = 4	Ele: 7.9%, n = 3 Mid: 3.6%, n = 2 High: 10.7%, n = 18 College: 6.2%, n = 18 Grad: 0%
Tourism experience $p = 0.008$	No: 47.9%, n = 186 Yes: 35.6%, n = 52	No: 43.8%, n = 170 Yes: 58.9%, n = 86	No: 8.2%, n = 32 Yes: 5.5%, n = 8
Marital status $p = 0.000$	Unmarried: 65.8%, n = 73 Married: 37.8%, n = 168	Unmarried: 27.9%, n = 31 Married: 54.6%, n = 243	Unmarried: 6.3%, n = 7 Married: 7.6%, n = 34
Age	$F(2, 543) = 33.317, p = 0.000$ *Post-Hoc Test* Mean difference: C1–C2 = –9.42, SD = 1.15, $p = 0.000$ Mean difference: C1–C3 = –4.91, SD = 2.27, $p = 0.078$ Mean difference: C2–C3 = 4.51, SD = 2.24, $p = 0.111$		
Length of residence	$F(2, 551) = 9.686, p = 0.000$ *Post-Hoc Test* Mean difference: C1–C2 = –6.30, SD = 1.47, $p = 0.000$ Mean difference: C1–C3 = –0.67, SD = 2.80, $p = 0.97$ Mean difference: C2–C3 = 5.63, SD = 2.77, $p = 0.107$		
Income	$F(2, 484) = 0.371, p = 0.690$		

supporters' (2.6%), though most are 'neutrals' (89.5%). People who have high school or college degree are more supportive of tourism preferences than middle school and elementary school. Interestingly, people who have graduate education become more neutral (66.7%) than supportive (33.3%).

Tourism experience appears to be a significant determinant against the three clusters. People who have *no* tourism experience are skewed toward either 'neutrals' (47.9%) or 'supporters' (43.8%) while those who have experience are more 'neutrals' (58.9%) than 'supporters' (35.6%). The interesting result is that, regardless of tourism experience, people are not opposed to tourism (no = 8.2% and yes = 5.5%). This finding indicates that people in the study area believe tourism to be a major income source, given the fact that tourism in Gross Island Product (GIP) accounts for about 20% (Jeju, 2014).

Marital status shows an interesting result, because the unmarried group is more 'supporters' (65.8%) than 'neutrals' (27.9%) and 'opponents' (6.3%), while the married group becomes more 'neutral' (54.6%) than 'supporters' (37.8%) and 'opponents' (7.6%). This may be because the unmarried group of people tend to be young adults who continuously seek out their permanent occupation. They find their temporary job easily in the tourism industry. However, the married group of people is relatively middle-aged and economically settled, wanting to spend more time with family members besides their job-related activities.

Age, length of residence and household income were analysed using ANOVA with post-hoc test. Age and length of residence are found to be significant against the three clusters while income appears to be insignificant. The significant difference is found that the younger group of people is more 'supporters' than 'neutrals' (mean difference C1–C2 = –9.42, SD = 1.15, $p = 0.000$). Similarly, length of residence shows that newcomers to the community are more 'supporters' than 'neutrals' (mean difference C1–C2 = –6.30, SD = 1.47, $p = 0.000$). The two findings of age and length of residence are interrelated, because people who have lived in the study area for a shorter period of time are young adults in comparison with old residents.

10.2.3 Wind farm development and clusters

Stakeholders' perceptions regarding wind farm development were measured using 22 items. In order to examine the relation between clusters and wind farm development, wind farm development was analysed using principal component factor analysis. All factor items with an eigen-value greater than one and factor loadings of 0.4 or higher were retained. Four factors were generated as follows: economic cultural impact, local impact, social impact and environmental impact. All four factors account for 66.04% of total variance and reliability of each factor was measured using Chronbach's alpha, resulted in an acceptable range from 0.741 to 0.922 (Table 10.3).

With factors associated with wind farm development, ANOVA was employed to examine if a difference exists between factors and the three clusters (Table 10.3). All four factors are found to be significantly different (economic cultural impact, ($F = 10.204$, $p = 0.000$); local impact ($F = 36.236$, $p = 0.000$); social impact ($F = 4.264$, $p = 0.015$); environmental impact ($F = 8.176$, $p = 0.000$)). An additional analysis was employed using MANOVA, which examines how individual items differ against the three clusters. A MANOVA analysis reveals that all items pertaining to wind

Table 10.3. Wind farm development and clusters.[a]

Wind farm development perceptions	Reliability	Sig. (ANOVA)	Tourism supporters	Tourism neutrals	Tourism opponents	Sig. (MANOVA)
Economic cultural impact	0.922	0.000 (F = 9.328)				
6. Wind farms are likely to result in increasing quality of education.			2.65	3.12	2.30	0.000
7. Wind farms are likely to create more jobs for community.			2.68	3.13	2.30	0.000
8. Wind farms are likely to provide more business for local people and small businesses.			2.82	3.17	2.09	0.000
5. Wind farms are likely to increase resident's income.			2.88	3.26	2.57	0.000
9. Wind farms are likely to improve local residents' overall quality of life			2.71	3.15	2.12	0.000
13. Wind farms are likely to promote more diversity of cultural activities.			2.58	3.03	2.24	0.000
14. Wind farms are likely to provide more parks and other recreational areas for local residents.			2.48	2.93	2.24	0.000
12. Wind farms are likely to provide an incentive for the preservation of the local culture.			2.29	2.70	2.42	0.000
Local impact	0.889	0.000 (F = 33.130)				
20. Wind farms are likely to improve local infrastructures such as water supply, drainage and power, etc.			3.31	3.72	2.18	0.000
21. Wind farms are likely to help to maintain roads and other public facilities.			2.94	3.56	2.15	0.000
19. Wind farms are likely to result in improving understanding/image of our community.			2.77	3.41	2.15	0.000
18. Wind farms are likely to increase additional tax revenue of local residents.			2.68	3.22	2.24	0.000

continued

Table 10.3. *Continued*

Wind farm development perceptions	Reliability	Sig. (ANOVA)	Tourism supporters	Tourism neutrals	Tourism opponents	Sig. (MANOVA)
22. Wind farms are likely to result in improved transportation services in the community.			2.54	3.26	2.52	0.000
15. Wind farms are likely to result in more cultural exchange between tourists and residents.			2.80	3.25	2.24	0.000
Social impact	0.741	0.018 (F = 4.033)				
16. Wind farms are likely to increase the prices of goods and services.			2.41	2.31	2.58	0.224
11. Wind farms are likely to increase crime rate.			2.06	1.92	1.90	0.261
17. Wind farms are likely to increase the value of property positively.			2.40	2.60	2.24	0.068
4. Wind farms are likely to result in traffic congestion.			2.26	2.02	1.93	0.008
10. Wind farms are likely to result in water pollution.			2.59	2.45	2.66	0.215
Environmental impact	0.829	0.000 (F = 8.006)				
1. Wind farms are likely to destruct local nature and wildlife.			2.91	2.55	2.46	0.001
2. Wind farms will destroy local uniqueness and landscape.			2.99	2.63	2.78	0.004
3. Wind farms are likely to increase noise and pollution.			3.17	2.86	2.85	0.006

[a] 1 being strongly disagree and 5 being strongly agree

farm development except for two items (increase the prices of goods and services and increase crime rate) from social impact are found to be significantly different. Overall, 'tourism neutrals' group is more positive toward wind farm development than the other two clusters. In particular, the 'tourism neutrals' group believes that wind farm development would bring some positive economic cultural impact and local impact to the community. Positively perceived impact such as economic cultural impact and local impact are seen in the group of 'tourism neutrals' only.

While the cluster 'tourism neutrals' group is found to be significantly different over the other two groups against two factors (economic cultural impact and local impact) and its difference is obviously seen in the 'tourism neutrals' group over 'tourism supporters' and 'tourism opponents' groups, the other two factors (social impact and environmental impact) found to be significant seem to be overly complicated. The social factor appears to be significantly different against the three clusters; only one single item under the social impact (traffic congestion) turns out to be significantly different at the level of 0.05, though all three clusters perceive traffic congestion not to be a result of wind farm development. Individual items under the environmental impact remain to be significant over the three clusters. The group 'tourism supporters' believes that wind farm development would increase noise and pollution while the other two groups disagree.

T-test was employed to examine if any difference between residents and government officials exists regarding wind farm development (Table 10.4). In the similar application, t-test was used twice over factors and individual items. The two stakeholders have mixed perceptions. With regard to economic cultural impact and local impact, the two stakeholders present significantly different opinions, i.e. the two factors are more positively perceived by the residents than by the government officials. In the meantime, differences between the stakeholders regarding social and environmental impact are not found to be significant. In other words, the two stakeholders are congruent in a consistent direction, either positively or negatively.

Individual items under the two factors – economic cultural and local impact – are also found to be significantly different except for one item (improve local infrastructure), though both perceive it positively. Individual items under social and environmental impact remain to be insignificant except for one each item ((increase the value of property – social impact) and (increase noise and pollution – environmental impact)). The level of disagreement of the two items is seen more strongly in government officials than residents.

Another t-test was conducted regarding tourism preferences. In the analysis, residents show more positive perceptions than government officials though both groups are interested in promoting diverse tourism-related preferences (Table 10.5).

In the study area, as mentioned above, the island is heavily dependent upon power imported from the mainland and is exposed to the risk of power shortages accordingly. Development of alternative energy sources, thus, may be inevitable to lessen potential power shortages. Several wind turbines were built and there are proposals to build more in the near future. Stakeholders pertaining to wind farm development must agree to minimize any possible conflict. In that regard, two stakeholders were asked to respond to the level of agreement regarding wind farm development (Table 10.5). The level of agreement is related with two questions: 'I supported wind farms development' and 'I support wind farms development'. The

Table 10.4. Comparison of wind farm development between residents and government officials.[a]

Wind farm development perceptions	Composite mean					
	Res	Gov	T-test	Res	Gov	T-test
Economic cultural impact	2.89	2.53	0.000 (F = 3.941)			
6. Wind farms are likely to result in increasing quality of education.				2.93	2.54	0.00
7. Wind farms are likely to create more jobs for community.				2.93	2.68	0.03
8. Wind farms are likely to provide more business for local people and small businesses.				3.01	2.64	0.00
5. Wind farms are likely to increase residents' income.				3.08	2.81	0.02
9. Wind farms are likely to improve local residents' overall quality of life				2.96	2.55	0.00
13. Wind farms are likely to promote more diversity of cultural activities.				2.87	2.33	0.00
14. Wind farms are likely to provide more parks and other recreational areas for local residents.				2.74	2.39	0.00
12. Wind farms are likely to provide an incentive for the preservation of the local culture.				2.58	2.24	0.00
Local impact	3.17	2.78	0.000 (F = 6.687)			
20. Wind farms are likely to improve local infrastructures such as water supply, drainage and power, etc.				3.49	3.31	0.12
21. Wind farms are likely to help to maintain roads and other public facilities.				3.23	3.03	0.12
19. Wind farms are likely to result in improving understanding/ image of our community.				3.14	2.70	0.00
18. Wind farms are likely to increase additional tax revenue of local residents.				3.04	2.45	0.00

continued

Table 10.4. *Continued*

Wind farm development perceptions	Composite mean					
	Res	Gov	T-test	Res	Gov	T-test
22. Wind farms are likely to result in improved transportation services in the community.				3.00	2.45	0.00
15. Wind farms are likely to result in more cultural exchange between tourists and residents.				3.07	2.74	0.00
Social impact	2.31	2.23	0.294 (F = 4.114)			
16. Wind farms are likely to increase the prices of goods and services.				2.38	2.22	0.15
11. Wind farms are likely to increase crime rate.				1.98	1.97	0.86
17. Wind farms are likely to increase the value of property positively.				2.55	2.21	0.01
4. Wind farms are likely to result in traffic congestion.				2.08	2.19	0.23
10. Wind farms are likely to result in water pollution.				2.52	2.50	0.81
Environmental impact	2.79	2.98	0.059 (F = 3.687)			
1. Wind farms are likely to destruct local nature and wildlife.				2.67	2.79	0.33
2. Wind farms will destroy local uniqueness and landscape.				2.73	2.92	0.13
3. Wind farms are likely to increase noise and pollution.				2.96	3.21	0.03

[a]1 being strongly disagree and 5 being strongly agree

Res, residents; Gov, government

result is interesting that regarding whether stakeholders '*supported*' wind farms development or not, government officials were negative. In other words, government officials did not support wind farm development. However, the question whether stakeholders '*support*' wind farm development reveals a different result. In particular, government officials who have observed wind farm development changed their negative perceptions to positive ones. This implies that government officials may have benefited, for example by economic growth and improvement of local infrastructure, from the wind farm development. Perceptions of residents regarding the development have also increased.

Table 10.5. Comparison of tourism preferences and wind farms support between residents and government officials.[a]

Items. I would like to see more . . .	Resident	Government	t	Sig.
families to visit	4.11	3.89	2.054	0.041
free independent travellers/backpackers to visit	4.15	3.92	2.232	0.027
packaged tourists to visit	4.09	3.71	3.345	0.001
diverse tourists to visit	4.22	3.99	2.232	0.026
tourism development	4.04	3.57	4.076	0.000
international tourists to visit	4.23	3.97	2.549	0.012
local residents' involved in tourism	3.98	3.53	3.784	0.000
Wind farms support				
I supported wind farm development	4.37	3.51	2.88	0.00
I support wind farm development	4.76	3.71	3.15	0.00

[a]1 being strongly disagree and 5 being strongly agree

10.3 Discussion

It appears that alternative energy source development may be inevitable where the power is not independently produced. It is essential that when planning large-scale development at a tourism destination, detailed information and attention needs to be given to stakeholders for further agreement and minimization of potential conflicts.

From the marketing perspective, understanding stakeholders' needs and wants is important in order to identify and develop specific target strategies. In the present study area, there has been long ongoing discussion with regard to the development and need for wind farms. While the debate is still in progress, some have been built and more are proposed at a tourism destination. In order to develop marketing strategies, a cluster analysis is common statistical method used for market segmentation (Dolnicar, 2003). Three clusters (tourism supporters, tourism neutrals and tourism opponents) from tourism development were obtained. Each cluster has distinct characteristics.

The group of 'tourism supporters' – highly educated and not married – appears more strongly in government officials. In comparison with other cluster groups, 'tourism supporters' does not seem to be supportive of wind farm development. Marketing strategies for this group may need to establish the perception that wind farm development can be diversifying tourism attractions, not solely focusing on alternative energy. For that purpose, it is suggested that studying previous case studies that have similarly underlined how wind farms attract tourists can be helpful. For example, around the world, such as in Scotland, Ireland, New Zealand and California, people pay to visit wind farms, which can increase local revenue (Pavlides, 2005; Toke, 2005). In addition, it is important to stress that, by establishing small turbines, wind farms do not produce as much noise as imagined (Devine-Wright, 2005; Warren *et al.*, 2005).

The second cluster group, 'tourism neutrals', has shown interesting results. More than half of the residents fall into this group with more females, high school diploma and married. Interestingly, people in this group have worked in the tourism industry much more than the other two cluster groups but are not much supportive of tourism preferences though still positive of them. This group is very interested in the economic, cultural and local impact that wind farm development would bring. Compared with the other two clusters, this group sees the development as a way to improve economic cultural opportunities such as quality of education, more jobs, income and recreational opportunities. This group also expects that wind farms improve local impacts such as local infrastructure and maintenance of public facilities. This finding is similar to the study conducted by Lee *et al.* (2013), who studied non-tourism development in a nearby area. In particular, residents in this group expect those aforementioned local impacts to be improved as a result of wind farm development. Local residents' voices often remain to be unheard, thus resulting in conflicts between stakeholders (Davis and Morais, 2004). Thus marketing strategies for this group are suggested that keeping local infrastructure up-to-date can be an important approach to embrace local voices by establishing a task force to ask questions, for example, such as '*Does the community need educational institutions?*' and '*How often are roads, water supply, drainage and power system being checked?*' on a regular basis.

The third group 'tourism opponents' sees tourism negatively as implied. People in this group are small samples (7.4%, n = 42/567 clustered samples). Compared to 'tourism supporters', people are relatively less educated, and are not supportive of tourism preferences regardless of their age, length of residence and household income. Similarly, this group seems to be very negative of any possible positive effects, in terms of economic and local impact, from the development of a wind farm. However, this group is not always negative of wind farm development; rather this group perceives any possible negative impact such as crime rate and traffic congestion to occur in the least negative way. The group of people in this cluster may want to stay where they are accustomed to. Developing a marketing strategy for this group may be difficult, but identifying what people in the group want and need can be a first step to embrace their voices. For example, questions such as 'under what circumstances do they feel benefited?' or 'have their voices been unheard?', 'do they feel happy if they receive tangible benefits?' may need to be advanced.

For this research two theories are used: stakeholder and congruence theory. First, stakeholder theory aims values in managing a firm. In the tourism industry, the theory is used widely to signify various voices, including residents, governmental bodies and entrepreneurs (Perdue, 2004; Bryd, 2007; Timur, 2010). Two stakeholders – residents and government officials – in this study represent each party. The results reveal that each stakeholder demonstrates different interests regarding tourism preferences and wind farm development. Residents in the study demonstrate more interest in any type of tourism development than government officials, indicating that residents have witnessed the importance of tourism and/or expect tourism to bring more benefits to the community.

The significant difference between residents and government officials is also seen in some factors under the wind farm development. In particular, perceptions of local impact such as improvement of local image and public transportation services are more positively identified in the resident group than government officials. Looking at composite mean of economic cultural impact, both stakeholders perceive wind farm development as not bringing many economic cultural benefits, but the significant difference is found to be stronger in the group of residents than government officials in regard to economic-related variables such as business opportunities, household income and additional tax revenue.

Second, congruence theory is used to examine how two stakeholders feel about tourism preferences and wind farm development. In general, if there are two contradicting people, individuals who feel pressure often change his/her judgement to the other side. However, if the two groups are similar, then there will be no problem and he/she will not experience pressure of any form. In terms of tourism preferences, both parties are supportive of them though the degree of agreement is different. In addition, even though the two parties feel similar toward social and environmental impact to be generated from wind farm development, the significant difference lies in two factors: economic cultural impact and local impact. Residents characterize wind farm development positively as a way to bring benefits such as more business opportunities, household income and improvement of local image while government officials are often negative of them. Thus, marketing strategies for each stakeholder need to be taken into consideration.

Government officials, for example, need to increase awareness of the importance of tourism so that residents feel on the same track with government officials. A specially designed programme pertaining to the importance of tourism may need to be developed for government officials so that they may make a greater effort to serve residents better.

10.4 Conclusions

An alternative energy source development is a popular topic over both academic disciplines and industries due to various reasons such as increased energy consumption and using up natural resources. Wind farm development has been considered to respond to an alternative source of energy. Despite the fact that wind farm development is regarded as an alternative source of energy, stakeholders around the development demonstrate different interests. Thus, this study examines how stakeholders perceive the development taking place at a tourism destination and what factors from the development stakeholders demonstrate similar or dissimilar agreement.

Interesting results emerge from the analysis to preferred tourism development and wind farm development. Although government officials are less convinced of their merits regarding tourism preferences, both stakeholders nevertheless are in favour of tourism preferences. The findings indicate that both stakeholders believe tourism to be a major economic entity on the island.

In regard to wind farm development, both parties demonstrate mixed perceptions. The notable disparity between residents and government officials indicates that residents portray economic cultural and local impact to wind farm development while government officials are less convinced of both impacts. Some of the arguments found to be significantly different include that residents expect wind farm development to increase business opportunities, household income and improve images of the local community while government officials are somewhat negative of them.

Despite the difference of opinions, to a certain extent, to wind farm development, both stakeholders agree with the necessity of wind farms. Wind farms may come to diversify tourism attractions and play an important role in semi-substituting current power imported from the mainland. Given the nature of a new type of development at a tourism destination, opinions will remain to be divided over whether such a development would be acceptable to various stakeholders, but this research finds that both stakeholders agree that such a development would improve local infrastructure and facilities. Thus, the next step, if planning more in the near future, is to develop a strategic framework that leads to shorten the disparity between stakeholders with mutual respect.

References

AIM Research A/S (1993) Holdningsunderøgelse til vind energi. Report prepared for the Danish Wind Turbine Manufacturers. Cited in: Devine-Wright, P. (2005) Beyond NIMBYism: towards an integrated framework for understanding public perceptions of wind energy. *Wind Energy* 8, 125–139.

Becken, S., Simmons, D. and Frampton, C. (2003) Energy use associated with different travel choices. *Tourism Management* 24(3), 267–277.

Bramwell, B. and Lane, B. (2000) *Tourism Collaboration and Partnerships: Politics, Practice and Sustainability*. Channel View Publications, Clevedon, UK.

Bryd, E.T. (2007) Stakeholders in sustainable tourism development and their roles: applying stakeholder theory to sustainable tourism development. *Tourism Review* 62(2), 6–13.

Cavallaro, F. and Ciraolo, L. (2005) A multicriteria approach to evaluate wind energy plants on an Italian island. *Energy Policy* 33(2), 235–244.

Chang, T.C. and Huang, S. (2011) Reclaiming the city: waterfront development in Singapore. *Urban Studies* 48(10), 2085–2100.

Dalton, G.J., Lockington, D.A. and Baldock, T.E. (2008) Feasibility analysis of stand-alone renewable energy supply options for a large hotel. *Renewable Energy* 33(7), 1475–1490.

d'Angella, F. and Go, F.M. (2009) Tale of two cities' collaborative tourism marketing: towards a theory of destination stakeholder assessment. *Tourism Management* 30(3), 429–440.

Davis, J.S. and Morais, D.B. (2004) Factions and enclaves: small towns and socially unsustainable tourism development. *Journal of Travel Research* 43(1), 1–10.

Devine-Wright, P. (2005) Beyond NIMBYism: towards an integrated framework for understanding public perceptions of wind energy. *Wind Energy* 8(2), 125–139.

Diederichs, A., Grunkorn, T. and Nehls, G. (2008) Offishore wind farms-disturbance or attraction for harbor porpoises? First results of t-pod investigation in HORNS REV and NYSTED. *Proceedings of the ASCOBANS/ECS Workshop at the European Cetacean Society's 21st Annual Conference*, The Aquarium, San Sebastian, Spain, 21 April 2007, pp. 42–49.

Dincer, I. (2000) Renewable energy and sustainable development: a crucial view. *Renewable and Sustainable Energy Reviews* 4(2), 157–175.

Dolnicar, S. (2003) Using cluster analysis for market segmentation-typical misconceptions. *Australasian Journal of Market Research* 11(2), 5–2.

Duic, N. and Carvalho, M. d G. (2004) Increasing renewable energy sources in island energy supply: case study Porto Santo. *Renewable and Sustainable Energy Reviews* 8(4), 383–399.

Eckstein, H. (1997) *Congruence Theory Explained*. Centre for the Study of Democracy, University of California, Irvine, California.

Healey, P. (1997) Strategies, Processes and Plans. In: *Collaborative Planning: shaping places in fragmented societies*. UBC Press, Vancouver, Canada, pp. 243–283.

Jamal, T. and Getz, D. (1995) Collaboration theory and community tourism planning. *Annals of Tourism Research* 22(1), 186–204.

Jeju (2014) Jeju statistics. Available at: http://www.jeju.go.kr/contents/index.php?mid=010905 (accessed 13 January 2014).

Jensen, M.C. (2001) Value maximization, stakeholder theory, and the corporate objective function. *European Financial Management* 7(3), 297–317.

Kaplanidou, K. and Vogt, C. (2003) Destination branding: concept and measurement. *Working Paper Michigan State University* – Michigan State University, Department of Park, Recreation and Tourism Resources, August.

Kempton, W., Firestone, J., Lilley, J., Rouleau, T. and Whitaker, P. (2005) The offshore wind power debate: views from Cape Cod. *Coastal Management* 33, 119–149.

Lee, S., Kline, C. and Lee, G. (2013) Perceived impact of a naval base on a tourism destination. *Asia Pacific Journal of Tourism Research* (online first). Available at: http://www.tandfonline.com/doi/abs/10.1080/10941665.2013.806937 (accessed 16 January 2014).

Lee, T., Wren, B. and Hickman, M. (1989) Public response to the siting and operation of wind turbines. *Wind Engineering* 13, 188–195.

Merrilees, B., Miller, D. and Herington, C. (2012) Multiple stakeholders and multiple city brand meanings. *European Journal of Marketing* 46(7/8), 1032–1047.

Osgood, C. and Tannenbaum, P. (1955) The principle of congruity in the prediction of attitude change. *Psychology Review* 62, 42–55.

Pavlides, L. (2005) The aesthetics of wind power. *Providence Journal*, 5 March.

Perdue, R. (2004) Stakeholder analysis in Colorado ski resort communities. *Tourism Analysis* 8(2–4), 233–236.

Phillips, R. (2004) Some key questions about stakeholder theory. *Ivey Business Journal*. Available at: http://iveybusinessjournal.com/topics/the-workplace/some-key-questions-about-stakeholder-theory (accessed 16 January 2014).

Robb, D. (2003) Blowin' in the wind. *Power Engineering*, April, 40–44.

Tabatchnaia-Tamirisa, N., Loke, M.K., Leung, P. and Tucker, K.A. (1997) Energy and tourism in Hawaii. *Annals of Tourism Research* 24(2), 390–401.

Toke, D. (2005) Explaining wind power planning outcomes: some findings from a study in England and Wales. *Energy Policy* 33, 1527–1539.

Timur, S. (2010) Analyzing urban tourism stakeholder relationships: a network perspective. University of Calgary, Canada. Available at: http://citeseerx.ist.psu.edu/viewdoc/download?doi=10.1.1.110.2329&rep=rep1&type=pdf (accessed 16 January 2014).

Warren, C.R., Lumsden, C., O'Dowd, S. and Birnie, R.V. (2005) 'Green on green': public perceptions of wind power of Scotland and Ireland. *Journal of Environmental Planning and Management* 48(6), 853–875.

Wolsink, M. (2000) Wind power and the NIMBY-myth: institutional capacity and the limited significance of public support. *Renewable Energy* 21, 49–64.

Wordwiz (2013) Trust power: Dundonnell wind farm EES consultation plan. Available at: http://www.dtpli.vic.gov.au/__data/assets/pdf_file/0003/230763/Dundonnell-EES-Consultation-Plan_May-2013.pdf (accessed 14 November 2013).

Part IV Challenges and Future Research Directions

11 Island Tourism: Challenges and Future Research Directions

Muzaffer Uysal[1]* and Patrizia Modica[2]

[1]University of Massachusetts, Amherst, USA; [2]University of Cagliari, Sardinia, Italy

11.1 Introduction

Tourism consumption usually takes place in physical settings. Islands as physical settings are also popular destinations for millions of travellers. However, the level of development, tourism infrastructure and accessibility, resource base and governance may show a great degree of variation from island to island. Each island destination is unique and has its own challenges and developmental opportunities (e.g. Jayawardena, 2005; Cave and Brown, 2012; Meyer-Arendt and Lew, 2013; Croes, 2016). 'The challenge to sustainable management is to keep those unique elements that are part of the mystique that attracts tourists' (Sheldon, 2005, p. 10). Thus, in this book we do not offer a definition for islands, rather, we let our contributors contextualize their choice of islands from the perspective of their research approaches. We had no priority conditions set on the scope of the chapters. However, our main goal was to create an edited volume that could bring together the interplays of sustainability, competitiveness and quality of life of stakeholders in the context of islands as destinations. As the field of tourism displays maturity and scientific sophistication, it is important that we as tourism researchers fully understand and further examine the breadth and depth of the salient aspects of sustainable island tourism issues such as policy matters, public and private interface, competitiveness and quality of life of stakeholders that engage in production of tourism goods and services and planning and development decisions.

There is a plethora of books with conceptual and empirical case-based studies that have exclusively examined sustainable tourism development and related policy issues relevant to islands as destinations (e.g. Conlin and Baum, 1995; Briguglio *et al.*, 1996a; Ioannides *et al.*, 2001; Apostolopoulos and Gayle, 2002; Gayle, 2002; Harrison, 2003; Duval, 2004; MacLeod, 2004; Belle and Bramwell, 2005; Butler and Hinch, 2007; Carlsen and Butler, 2011; Hampton and Jeyacheya, 2013; Stratigea and Katsoni, 2015; Croes and Rivera, 2016). Islands as destinations are

*E-mail: muzzo.uysal@umass.edu

sometimes referred as micro-states or small island developing states (SIDS). Some of these studies used case study approaches (Briguglio *et al.*, 1996b) and others conducted empirical studies to focus on a given tourism issue such as development, poverty reduction, growing complexities in sustainable island tourism development, sustainable tourism policy, economic impacts, or quality of life issues and corporate social responsibility (e.g. Kokkranikal *et al.*, 2003; Tsartas, 2003; Chen, 2006; Farsari *et al.*, 2007; Sahli *et al.*, 2007; Dodds and Joppe, 2009; Hamzah and Hampton, 2013; Croes and Rivera, 2016). The notion of sustainability and sustainable tourism development, including ecotourism, is one of the research topics that occupied the scope and content of such examinations (Carlsen and Butler, 2011). Because of their uniqueness as destinations, competitiveness was also included in such studies (Buhalis, 1999). However, although the interplays of sustainability, competitiveness and quality of life were implied, studies that examine the interplays of these constructs in general, and quality of life issues have received rather limited attention. It is clear that the elements of sustainability and the constructs of competitiveness and quality of life are intertwined and intimately connected, suggesting further concerted research efforts.

11.2 Challenges and Issues

A number of previous studies also identified a wide variety of island features and vulnerability elements, some of which are posed by externalities to the system of island destinations. MacLeod (2004) and other researchers such as Choi (1992), Briguglio (1995), Conlin and Baum (1996), Ioannides *et al.* (2001), Kim and Uysal (2002), Baldacchino (2006), McElroy (2006), Scheyvens and Momsen (2008), Brown and Cave (2010), Cuccia and Rizzo (2011), Hampton and Jeyacheya (2013), Sarrasin (2013) and Su *et al.* (2016) all mentioned common factors that are shared by (small) island destinations. These factors include limited resource base, limited economies, dependency on dominant tourism sectors, separateness, small scale, isolation, proximity to origin market, accessibility, seasonality, and difference in terms of political structure, climate, biodiversity and environment, infrastructure, and their vulnerability to market forces, hence, dependency. MacLeod (2004) further articulated that problems faced by islands are commonly attributed to finite resources that impinge on the sustainability of the tourism economy. Some island destinations are yet to utilize their resources effectively. This point is also echoed by a countless number of studies examining island destinations (e.g. Butler and Stiakaki, 2001; Sharpley, 2001; Hampton and Christensen, 2007: Rodriguez *et al.*, 2008; Michael Hall, 2010; Sarrasin, 2013; Chen *et al.*, 2016). There is tremendous potential here for researchers to connect use of resources to the well-being of locals and destination competitiveness in island destinations (Croes, 2016). The question is then how do islands as destinations continue to develop tourism in a way that is sustainable, remains competitive and also improves the well-being of both residents and providers of tourism goods and services? Understanding the interplays of such a combination would enable island destination management to maintain and increase their market share and also meet their social obligations to their community and local governments.

Although we have seen a sustained level of research attention devoted to island and micro-state destinations, the uniqueness of these places as destinations pose new challenges and opportunities for both tourism academics and professionals in today's highly competitive and globalized international tourism market. Over the past three decades or so we have also witnessed a higher level of interest in managing and dealing with crises and risk management issues associated with both natural disasters and socio-political instability. The main area of research mostly centres on policy orientations, guidelines, public/private sector planning partnerships; product development and diversification; marketing; human resource management; well-being of stakeholders and sustainability. Performance measures/indicators and monitoring activity use-level and change over time are also imperative in the same vein of research unique to island destinations. We encourage researchers not only to develop appropriate and island-contextualized indicators but also link such indicators to measurable management outcome variables such as satisfaction and well-being of stakeholders. This point has also being brought to the forefront of our research agenda for island tourism research by a recent special issue of the *Journal of Destination Marketing and Management* (vol. 5, 2016), guest-edited by Prof. Robertico Croes.[1] The special issue explores the interplays and connectedness of three constructs: tourism development, small island destinations and the well-being of the island residents. Croes (2016, pp. 3–4) highlights three lessons from the special issue: (i) the relationship between income and well-being – happiness of residents in small island destinations is not self-evident; (ii) the level of income effects that shape residents' well-being depends on meditating factors such as social compassion and the degree of concurrence of interest among stakeholders; and (iii) residents' well-being may be explained by the nature of the relationship between government officials and residents. These propositions and empirical substantiations independently and collectively signal the complexity of island-specific issues and call for further research and case studies reflective of the uniqueness of island tourism destinations. We believe that this line of research is making its way into our mainstream tourism research and will receive heightened interest in the years to come.

Sustainability performance indicators unique to island destinations are an important tool for destination development and management (Reddy, 2008; UN, 2014). The development of relevant indicators can serve as a valuable tool in identifying types of tourism product development and what aspects or factors influence tourists in their decision to visit other countries (Dwyer and Kim, 2003). These indicators are defined as sets of information used to measure changes in specific aspects of tourism destinations and operations (Dupeyras and MacCallum, 2013; Modica, 2015; WEF, 2015; European Commission, 2016). Whether they take the form of qualitative or quantitative measurements, they are meant to provide reference points for business decisions. Croes (2011, p. 440) advocates that small island destinations are well served in using performance indicators because they provide guidelines, they correct for inefficient management directions, and they promote positive effects of competition among destinations. Performance indicators should be reviewed on a regular basis and updated with the most recent information in order to be relevant for decision making and also allow public authorities and operators to review previous data and assess subsequent needs and objectives for the following years (Tanguay *et al.*, 2013).

In island destination development, the procedure for the development and implementation of indicators should have a structure reflective of the destination design and location characteristics. First, the research and organization stage should define the destination and its vision, identify its important tourism assets and evaluate the situation at hand. Then, the indicators are actually developed. Based on the selected issues such as environmental impacts of tourism development, quality of life of stakeholders and destination competitiveness at hand, indicators are calculated and established. The third stage is the implementation of indicators. The indicators are evaluated to decide what new processes should be established in order to make improvements. More data are generated over time as these procedures are put into place, and based on the data, corrections and revisions can be made in hopes to lead to improvement.

The United Nations World Tourism Organization (UNWTO, 2004) provides over 50 issue areas for which indicators may be developed. The issues cover such areas as:

- Socio-cultural (issues related to community well-being, cultural assets, community participation, tourist satisfaction);
- Economic (capture of benefits, sustaining the tourist product, seasonality, leakages);
- Environmental (protection of valuable natural assets, managing environment resources: water, energy, waste);
- Tourism planning and management (destination planning and design of products and services, controlling use intensity, transport, marketing, branding); and
- Responding to global issues (e.g. climate change, epidemics, sex tourism).

Related to this discussion is that indicators to be launched should respond to special areas of interest identified for a given destination place. For example, socio-cultural indicators may cover issues related to community well-being, cultural assets and community participation. On the other hand, tourist satisfaction may have the following indicators: capture of benefits, sustaining the tourist product, seasonality and leakages. Environmental indicators may cover protection of valuable natural assets, managing environmental resources (water, energy, waste), while tourism planning and management's indicators include: destination planning and control, design of products and services, controlling use intensity, transport, marketing, branding as well as responding to global issues such as climate change, epidemics and sex tourism. The challenge here is to find the best combination of relevant issues reflective of the destination place at hand and operationalize them to fit the goals and implementation strategies of the destination community (Choi and Sirakaya, 2005).

Another issue area that should be added to the list is about objective and subjective well-being factors for the destination place and its stakeholders. Developing such indicators is important for destination development and management because it helps assess a destination's competitiveness and its use of resources, which is directly linked to the ability to perform better in the delivery of goods and services to tourists than other destinations. If the tourism agency for a destination is able to understand, based on indicators, where their destination may be falling short compared to competitors, then they will be able to create implementation plans to address these issues and make their destination more competitive.

Island destinations such the Caribbean with similar resources and amenities can be under pressure to develop and maintain a sustainable competitive advantage over the same destination competitive set, they must be extremely mindful of emerging trends, forecasts and future planning. If they do not prepare themselves, they run the risk of quickly falling behind, depleting their resources and even destroying the natural environment as well as the well-being of the indigenous people who live there year round (e.g. Duval, 2004; MacLeod, 2004; Jayawardena, 2005; Daye *et al.*, 2008). Obviously, the main goal is to attract as many tourists as possible and make a profit, but destination management officials must be wary of spoiling their chances to remain a viable long-term destination by not taking proper precautions.

It is clear that in island destinations the concepts of sustainability, environmental preservation and well-being are very important to maintaining a competitive advantage. Sustainability planning is key to ensuring the destination can keep up with demand. Decisions such as how many people to permit at a certain attraction can be made by studying indicators such as peak times, frequency of use and amount of waste disposal, among others. By analysing this type of data, managed attractions in destinations can create policies that set physical limits on how many people can be allowed in an area at a time. This is especially important for cultural attractions such as historical ruins that need to be monitored closely for signs of damage. By ensuring an attraction has a sustainable strategy, tourists are likely to have satisfying experiences, which in turn may affect visitation positively.

Environmental preservation is also crucial, especially in locations such as islands and micro-destinations with attractions that depend on the health of the physical environment. A prime example of environmental preservation leading to a competitive advantage at the destination is the Galapagos. The Galapagos Islands have very strict visitation policies because of the fragility and uniqueness of the area (Kenchington, 1989). There are rules to govern who can come to study the animals and plants and for how long. This allows them to preserve the natural beauty and educational opportunities the destination offers. Other examples include national parks and laws about remaining on trails in order to keep from disturbing the natural environment. Managing attractions and monitoring demand for them is of immense importance for both sustaining natural resources and providing satisfactory visitor experiences while improving the quality of life of stakeholders (Jensen, 2014).

Finally, well-being of the indigenous people as well as the tourists is important to monitor because destinations want to please visitors and keep them coming back without impinging on the well-being of locals as a result of tourism and recreation activities. Well-being of locals can be measured by a number of indicators such as availability of housing and ability to afford living in the area. Unfortunately, many locations neglect this concept and many locals become unhappy and sometimes displaced from their homes because they cannot afford to live there after the destination grows in attractiveness. Because island destinations have limited space and resources, it is not unlikely to see that indigenous people are being displaced as a results of higher property taxes, land and cost of living. One way to monitor this is to develop appropriate indicators to continue to measure both physical changes in the destination and behavioural response to these changes over time and analyse the ratio of tourists to locals in locations in order to gauge the physical, cultural and psychological aspects of carrying capacity of the destination (Manning *et al.*, 2002). The island tourism area life cycle needs to be assessed and monitored and longitudinal studies are needed (Hamzah and Hampton, 2013). By doing so, we can ensure a sustainable and competitive tourism business and appropriate responses to different phases of tourism development (Uysal *et al.*, 2012a). We challenge researchers and policy makers to engage in more creative and innovative ways of responding to growth and change so that there is resilience and stewardship that would propel sustainable tourism development.

Another key challenge about island tourism is related to its physical, social and psychological carrying capacity in figuring out limits of acceptable use and demand without placing more pressure on the ecosystem or impinging on the degree of competitiveness and well-being of its residents and relevant stakeholders. Rapid changes in demand as a result of low or high seasons could easily end up forcing overcapacity, insufficient use of infrastructure, reduction of workforce or lack of further investment during shoulder and low seasons (Kerstetter and Bricker, 2012; Ridderstaat *et al.*, 2014). However, the general consensus is that tourism has an overall positive impact on a destination, bringing in a significant amount of revenue and local jobs. On the other hand, as tourism increases in island destinations, carrying capacity may be reached and the structure and nature of visitor flows may change depending on seasonality, even causing a visible market loss or decline and thus less revenue and deterioration of quality of life (Manning *et al.*, 2002; Kerstetter and Bricker, 2012; Kim and Uysal, 2002). Island destinations are likely to be more vulnerable to visitation patterns, thus, monitoring demand and measuring structural and behavioural change over time is of immense importance to island destination management. Clearly, we need to do more research in this area to help develop specific tools and strategies that would offer flexibility and ease of use to deploy and implement. Partnership between different stakeholders and securing institutional support would certainly play a significant role as public policy can be established to prevent the destruction of certain historical and cultural sites and landmarks with high tourist volumes and mitigate the negative consequences of increased visitation. Also, local tourism conservation and special interest groups can be formed to help oversee impacts on a destination and preserve identity because such indigenous groups are familiar with the condition of the island destination before experiencing the influx of visitors. It is essential to maintain a network between locals and the government agencies to support development and improvements (Tolkach and King, 2015). Strategies and tools should be reflective of the phases of tourism development. For example, an island destination experiencing the consolidation/stagnation phase, as its limit of acceptable use is nearing the set or maximum threshold, should deploy conservation marketing techniques to inform and educate visitors and locals on the impacts that have already been encountered. Education and knowledge is an easy way to help mitigate some of the effects felt from large amounts of tourism. All in all, island destinations must be aware of where they are in the tourism life cycle in order to stay competitive and prevent a potential decline both in the quality of tourist experience and quality of stakeholders' life, if carrying capacity is not carefully monitored.

Destination competitiveness is a very complex phenomenon with multifaceted aspects. There are so many internal and external forces that influence competitiveness (Crouch, 2010; Crouch and Ritchie, 2012). Thus, there are many benefits of having good indicators to combat the complexity of destination competitiveness and management challenges. As indicated with general performance indicators, good indicators relevant to competitiveness allow for better decision-making, prevention of potential problems through the identification of emerging risks, and if a problem does arise, good indicators help with corrective action by helping to identify the impacts of a problem. They also act as a performance measure of the implementation of development plans and management actions. They also provides greater public accountability, better communication and constant monitoring, which allows for continuous improvement, although not every destination uses the same indicators or uses them in the same way. Thus, it is important to evaluate an indicator's feasibility: relevance to the issue, data availability (capacity to collect and process), credibility of the information, clarity and understandability to users, and comparability over time and across jurisdictions or regions.

There is enough evidence to suggest that current competitiveness measures or indicators may not respond to special needs of small island destinations and the traditional small island tourism development model, which is based on sun, sand and sea, may not always be the best way to sustain tourism activities and development (Craigwell, 2007; Croes and Semrad, 2013). Alternative and island-specific measures need to be developed and explored. For example, Croes (2011) presented an alternative measurement of a more suitable Tourism Competitiveness Index (TCI), and through the application of panel regression analysis, the model was able to track the most important factors such as size, dynamism and structure specifically related to a small island destination, affecting competitiveness among island destinations. The findings imply likelihood that more expensive destinations obtain a larger share of regional tourism revenues. Hence, providing a high-quality product with differentiated visitor market may allow destinations to become and remain competitive. The proposition certainly needs to be further tested in different island cases. We encourage researchers to seek out and conduct research with non-traditional product options (e.g. specialized recreactional activities, culture and traditions, gastronomy unique to indigenous ways of life) that could be augmented with traditional products such as sun, sand and sea of small island tourism destinations.

Prof. Robertico Croes, one of the leading scholars and experts in island economies, has done a number of empirical studies that support the notion that small islands can experience stronger basic patterns of growth than many developed countries, especially where economies of scale are not an issue, and advocates specialization for small islands in tourism (Croes, 2013). Based on 35 islands, of which 20 were small islands (population with less than 1.5 million), he found that size, a lack of complete independency and export orientation do not seem to affect the variance in the real per capita GDP at a greater degree, implying that the success of tourism specialization does not depend exclusively on comparative advantage and institutional realities, thus there is no harm in allocating resources to tourism development. He further points out that, unlike the common belief, empirical realities support that small islands have performed well in specializing as tourist destinations, despite their size constraints and resulting disadvantage in economic terms. The Caribbean region is an appropriate proxy for small island economies. However, globalization may be taking its toll on sectoral specialization in the Caribbean as a development strategy. He argues that the supply-side strategy seems to have lost its effectiveness. It is no longer providing differentiation advantages for competition. To reverse this trend, he advocates the implementation of new tourism strategies in the era of globalization based on a demand-focused approach to sustain value enhancement and sustained growth of tourism (Croes, 2006). There is ample opportunity to engage in demand-driven strategies to further develop island tourism and also investigate how this demand-driven growth affects the well-being of residents and allocation of resources and the role of governments.

11.3 Contribution of this Book

Our goal is not to cover every possible subject that may fall under the realm of island tourism but rather to offer a selection of topics that would exemplify the best work of our contributors and provide a portfolio of empirical, conceptual and case studies that would represent the issues of island tourism with respect to sustainability, competitiveness and quality-of-life issues. We believe that this book would be of great interest to the students of tourism, recreation, tourism planners and developers, decision makers and the like. In addition, practitioners and island destination management may find the book very useful in transferring and applying discussion points, suggestions and findings to their own unique island destination settings since the book focuses on the merits and importance of the interplays of sustainability, competitiveness and quality of life of island tourism and its stakeholders.

11.4 Overview of the Contents

Eighteen outstanding scholars representing several countries contributed their work to this book. We have invited those researchers who have the knowledge and expertise to share their work and field experience under the umbrella of the focus of the book, *Sustainable Island Tourism: Competitiveness and Quality of Life*. They infuse their passion into their writings when communicating and addressing the nature of sustainable island tourism and how the elements of competitiveness and quality of life of stakeholders are intimately connected to understand and examine challenges that island destinations are facing.

While introducing the chapters, we paraphrased summary points of the chapters to some extent and relied on what our contributors provided rather generously. With this acknowledgement, we introduce the chapters in the following section.

Part I: Sustainable Island Tourism

Islands as destinations also go through different phases of development, each requiring a set of management and marketing responses. The challenge is how to assess and develop the level of threshold that could place the island destination in equilibrium with respect to the consequences of tourism development, thus allowing destination management to monitor change over time and develop appropriate strategies. The first chapter, 'Sustainability and Tourism Development in Island Territories' by Patrizia Modica and Muzaffer Uysal, develops the concept of equilibrium in island destinations in the medium long-term perspective, based on Carrying Capacity and Limits of Acceptable Change – LAC – theories. The case study considers Sant'Antioco, a small Mediterranean island in the proximity of Sardinia, Italy and the potential of LAC in favouring consensus building for sustainable planning and policies, through the involvement of tourism and related sector entrepreneurs.

The second chapter, 'Ecotourism and Environmental Management: A Case Study of a Partnership for Conservation in Fiji' by Kelly S. Bricker and Deborah L. Kerstetter presents a case study of Rivers Fiji, an ecotourism company that established a lease for conservation of the Upper Navua River in Fiji. The authors review Fiji's current sustainability initiatives, including its Green Growth Framework and four multi-lateral environmental agreements that established and/or supported management of protected areas in Fiji. This is followed by discussion on strategies being used by stakeholders to manage protected territories and areas, and to what extent they are affected by customary governance practices. To ensure that other ecotourism companies can duplicate the process used by Rivers Fiji to establish a lease for conservation of the area surrounding the Upper Navua River, the authors note the unique structure of the lease, the steps taken to establish the conservation area as a Ramsar site, the company's educational outreach practices and the direct benefits and economic impact of establishing the lease. The chapter provides suggestions and discussions as to what other ecotourism researchers, companies and island countries/states can learn from the Rivers Fiji case study.

There is no question that the fragile nature of island tourism with its inherited limitations and unique characteristics poses challenges and calls for the attention of planners, decision-makers and academicians with regard to the developmental pattern of tourism in island destinations. Of particular importance is the need for islands to improve the quality of life of the local community, minimize seasonality and enhance their competitiveness.

It is noted that despite the attempts to identify indicators for the implementation of sustainable tourism, islands are challenged in their ability to transfer the rhetoric of sustainability into reality. The chapter 'Rhetoric versus the Realities of Sustainable Tourism: The Case of Cyprus' by Anna Farmaki, Levent Altinay and Muhammet Yaşarata presents two examples of sustainable projects implemented in Cyprus and discusses the way sustainable tourism implementation is inhibited by several factors including a limited understanding of the concept of sustainability and a general lack of sustainability awareness, the lack of governmental support in terms of the sustainable development of tourism, the poor coordination of activities between national and local stakeholders, a weak decision-making structure that marginalizes certain interested groups from policy design and the environmental constraints and limited employment opportunities present in island contexts.

Part II: Competitiveness and Quality of Life

There have been a significant number of studies that have focused on competitiveness and also developed models. The most commonly cited competitiveness model was developed by G.I. Crouch and B.J.R. Ritchie. This chapter, 'The Various Faces of Competitiveness in Tourism: A Survey of the Extant Literature' by Robertico Croes and Marketa Kubickova, provides a glimpse into the debate concerning tourism competitiveness by presenting a survey of the issues that inform the current debate on tourism competitiveness. The authors point out that the extant tourism competitiveness literature is rather fragmented and dispersed, and spans a wide range of tourist destinations and products. The result of this situation is that current research results do not make for easy comparison, and the existing knowledge on destination competitiveness is still scarce. The construct of competitiveness is complex and needs to be dissected in order to reveal a number of research pathways through which destination managers can gain more meaningful insights into the concept of tourism competitiveness. By doing so, one would hope that elements of competitiveness can be better contextualized to fit the needs of island destinations and their link to sustainability and quality of life could be operationalized.

A good example of developing appropriate competitiveness factors and presenting them in a meaningful way using a common tool called Importance/Performance Analysis (IPA) is undertaken by Jack Carlsen and Mariyam Zulfa in their chapter 'Island Tourism Competitiveness and Sustainability in the Maldives'. It begins with a synthesis of extant studies into a matrix of variables previously employed to measure competitiveness. Six groupings comprising 48 competitiveness factors were tested on a range of tourism experts in the Maldives to assess both the importance and performance of each variable. Using several IPA grids the chapter provides a summary of findings and a means of identifying and assessing the factors that make the Maldives competitive and that therefore should provide the focus of sustainable competitive advantage. This tool of IPA also allows areas where the Maldives could improve its competitive performance while setting priority areas that can be addressed in future policy, planning and management activities.

The chapter 'Quality of Life, the Tourism Area Life Cycle and Sustainability: A Case of Cuba' by Meghan Beardsley explores the potential interrelationship that may exist between quality of life, the tourism area life cycle (TALC) and sustainability for an island destination such as Cuba. Utilizing the case of Cuba, which is a destination in political, economic, cultural and tourism transition, the aforementioned constructs are identified as key elements that can explain the current state of affairs as well as predict future warning signs. As a result of the case, Cuban locals and inbound tourists alike must remain aware of threats to quality of life, tourism carrying capacity, and the threat to the derailment of the sustainable development track that the country is currently on. The rejuvenation phase of tourism development would certainly require partnership between stakeholders and active support from government entities.

It is clear that the factors that are likely to influence island sustainability and quality of life also have behavioural dimensions that need to be carefully conceptualized and contextualized. Chapter 7 'Place Satisfaction, Place Attachment and Quality of Life: Development of a Conceptual Framework for Island Destinations' by Haywantee Ramkissoon delves into how visitors' place-satisfaction contributes to overall quality of life, without which the sustainability of the tourism industry in small island economies may be threatened. Adopting the attitude-behaviour framework, the main objective of this chapter is to propose and develop a theoretical framework bridging the nexus between place satisfaction, place attachment, pro-environmental behaviour and overall quality of life of local visitors at distinctive cultural and natural attractions in island destinations. The integrated model draws from the environmental psychology literature proposing the mediating effects of: (i) place attachment on the relationship between local visitors' place-satisfaction and pro-environmental behaviour; and (ii) local visitors' pro-environmental behaviour on the relationship between place attachment and their overall quality of life.

Part III: Sustainability and Alternative Resource Use

Supply-oriented tourism strategies have well served the Caribbean destinations for a long time. However, in the light of a highly competitive and globalized international tourism market, new demand-driven and innovative strategies are needed to remain competitive and sustain growth. Chapter 8, 'Developing a Sustainable Caribbean

Tourism Product' by Allan Wright, attempts to develop a sustainable Caribbean tourism product. This is achieved through various steps. First, a description of the concept of sustainability, and particularly the multifaceted issue of sustainability in tourism, is explored. Then, the topics of competitive strategy and sustainability are examined, which leads to a thorough analysis of the concept of sustainable competitive advantage. An analysis of the literature review relating to these topics is undertaken. Afterwards, the strategies that the world's biggest companies have undertaken to develop a sustainable competitive advantage through their products are examined. These sections lay the basis for an analysis of the Caribbean, by doing so, the main policy recommendations for the Caribbean tourism industry are formulated. For an established destination region such as the Caribbean, new product initiatives and innovative strategies are deemed important for further growth in the global tourism market.

Some island destinations thrive on coastal and water-based activities for their main source of tourism income. These activities are propelled predominantly by local entrepreneurs and small businesses. The chapter 'Dive Tourism and the Entrepreneurial Process in the Perhentian Islands, Malaysia' by Julia Jeyacheya and Mark P. Hampton examines dive tourism in a small Malaysian island, particularly factors influencing the entrepreneurial process. The authors point out that much research on tourism entrepreneurs is derived from analysing business in the developed world, and has focused on the individual, not the process. Significantly less research exists for middle-income and less developed countries despite the critical role tourism plays in national development planning. The chapter contributes primarily to the initial knowledge gap by drawing upon extensive fieldwork interviews in the Perhentian Islands. The findings of this case study broadly reveal dive tourism as a low-cost industry for island economies that is transforming into a strategically competitive one. It is also revealed that the factors influencing the dive tourism entrepreneurial process are embedded within the social and historical context of Malaysia.

Islands naturally have limited resources, including energy to support development. As the energy consumption with traditional sources around the world increases, alternative energy resources have received great attention. In particular a destination where energy sources are heavily dependent upon external assistance deems to be interested in developing domestic energy resources. The involvement of multi-stakeholder groups in the development of large-scale infrastructure may bring some disagreement between stakeholders. Chapter 10, 'Stakeholders' Perceived Impact of Wind Farms at a Tourism Destination: A Case Study of Jeju Island, South Korea' by Seungwoo Lee, examines multi-stakeholders' perceptions and also their level of congruity on several aspects of wind farm development on Jeju Island. Gaining support from government authorities and establishing partnership between different stakeholders (tourism providers, residents and government entities) can make dependency, in this case on the mainland, less, hence, creating efficiency and sustainable tourism activities.

Part IV: Challenges and Future Research Directions

From the included chapters and preceding discussions it is evident that most small-island tourism destinations consider the tourism industry a major factor in ensuring their long-range socio-cultural and economic viability. They also face a number of challenges and also new opportunities in the globalized and highly competitive travel and tourism market. Through the adoption of tourism as a catalyst for economic advancement, island destinations hope to improve and sustain their regions' overall quality of life. If some degree of economic development is achieved, the overall communities' economic outputs may be augmented through individual labour earnings, increased value of property, and returns on capital investments in various tourism developments. Moreover, a wide variety of sustained visitor markets will enhance the financial viability of local businesses, such as accommodation, food services, retail stores, managed attractions and so on. Also, island governments benefit from tourism activities and tourists' expenditures in terms of fees and tax revenues. Ambitious and innovative demand-oriented tourism strategies would allow for small-island destinations to build upon the economic resources of tourism by diversifying their scope or providing further specialization in product offerings with targeted markets in mind (Bull and Weed, 1999; Buckley, 2002; Kim and Uysal, 2002; Veríssimo *et al.*, 2009; Lewis-Cameron and Roberts, 2010; Croes, 2011), while maintaining the level of quality of life desired in the destination. For these activities to result in positive outcomes, there needs to be a strong collaboration and formal and informal partnership among different stakeholders. An optimal sustainable state of island destinations should be able to satisfy both visitors and the local community (Lim and Cooper, 2009).

In order to maintain tourism as one of the valuable sources of economic and socio-cultural benefits to an island destination, island-tourism policy makers should consider not only short- and medium-term viability, but also the future, long-range strategic position of their touristic attractiveness. It is also explicit that tourism may not always uplift the economic well-being of a less-privileged portion of residents. The role of government is crucial in bringing a wide range of benefits both economic and non-economic such as more open green space, quality education and better health services to the poor as a result of tourism activities (Scheyvens, 2011). Furthermore, Craigwell (2007), in his study 'Tourism competitiveness in small island developing states' for the UN, reported that there is a need for small island countries to improve their levels of investment for the development of the technological infrastructure (telecommunications technology), and more importantly to ensure that the economic gains from tourism activities are also spread to the lower-income households. Regardless of the current level of tourism activities, an appropriate future formulation for sustaining a competitive position is key to maintaining a region's long-range economic and social well-being of its stakeholders.

The balance of attracting visitors to island destinations in a profitable manner while also enhancing the well-being of destination residents and goods and service providers, and preserving the natural and cultural assets of the destination for future generations is the most challenging aspect of developing any tourism destination. In this vein, tourism development in island destinations is not the exception. Furthermore, the interplays of sustainability, environmental preservation and quality of life can also be used to enhance destination competitiveness. The common link between these three constructs is the importance of natural and cultural assets to the future of the destination entity as a quality place to live as well as an attractive destination to visit (Boley and Perdue, 2012). Thus, the notions of sustainability, environmental preservation and quality of life should be important constructs for destination competitiveness not only because it would provide sustained economic benefits in the long run, but because it is the honourable and moral thing to do. There is also enough evidence to conclude that tourism has the leverage to be used as a tool to foster a better quality of life in the destination community (Uysal *et al.*, 2012b). There is ample opportunity to conduct research in this area. We wholeheartedly encourage researchers to further explore the interplays of destination competitiveness, sustainability and quality of life construct in island destination settings.

Note

[1] In the past there were some other special issues that were devoted to island and coastal destinations, for example, although not intended to be exhaustive, see: *Tourism* 49(4), 2001; *Tourism Geographies* 15(1), 2013; *International Journal of Culture, Tourism and Hospitality Research* 4(2), 2010; Asia *Pacific Journal of Tourism Research* 15(3), 2010; *Tourism Economics* 13(4), 2007; *Sustainable Development* 14(2), 2006; *International Journal of Contemporary Hospitality Management* 15(3), 2003.

Acknowledgements

A book like this would not have been possible without the generous and full support from our esteemed colleagues around the world. We thank all the contributors and are grateful for sharing their time, talent and expertise in writing their valuable chapters. We are also grateful for the support and encouragement of CABI and its wonderful staff members for helping to shape this book to reach its current form. Finally, we thank our family members for their constant support.

References

Apostolopoulos, Y. and Gayle, D.J. (2002) *Island Tourism and Sustainable Development: Caribbean, Pacific and Mediterranean Experiences*. Greenwood Publishing Group, Westport, Connecticut.

Baldacchino, G. (2006) Islands, island studies, island studies journal. *Island Studies Journal* 1(1), 3–18.

Belle, N. and Bramwell, B. (2005) Climate change and small island tourism: Policy maker and industry perspectives in Barbados. *Journal of Travel Research* 44(1), 32–41.

Boley, B.B. and Perdue, R.R. (2012) Destination management, competitiveness, and quality-of-life: a review of literature and research agenda. In: Uysal, M., Perdue, R. and Sirgy, M.J. (eds) *Handbook of Tourism and Quality-of-Life Research*. Springer, the Netherlands, pp. 515–528.

Briguglio, L. (1995) Small island developing states and their economic vulnerabilities. *World Development* 23(9), 1615–1632.

Briguglio, L., Archer, B., Jafari, J. and Walla, J. (1996a) *Sustainable Tourism in Islands and Small States: Issues and Policies*. Pinter, New York.

Briguglio, L., Butler, R., Harrison, D. and Filho, W.L. (1996b) *Sustainable Tourism in Islands and Small States: Case Studies*. Pinter, New York.

Brown, K.G. and Cave, J. (2010) Island tourism: marketing culture and heritage-editorial introduction to the special issue. *International Journal of Culture, Tourism and Hospitality Research* 4(2), 87–95.

Buckley, R. (2002) Surf tourism and sustainable development in Indo-Pacific Islands. I. The industry and the islands. *Journal of Sustainable Tourism* 10(5), 405–424.

Buhalis, D. (1999) Tourism on the Greek Islands: Issues of peripherality, competitiveness and development. *International Journal of Tourism Research* 1(5), 341.

Bull, C. and Weed, M. (1999) Niche markets and small island tourism: the development of sports tourism in Malta. *Managing Leisure* 4(3), 142–155.

Butler, R. and Hinch, T. (2007) *Tourism and indigenous peoples: Issues and implications.* Butterworth-Heinemann, Oxford, UK.

Butler, R. and Stiakaki, E. (2001) Tourism and sustainability in the Mediterranean: Issues and implications. In: Ioannides, D., Apostolopoulos, Y. and Sonmez, S. (eds) *Mediterranean Islands and Sustainable Tourism Development. Practices, Management and Polices.* Continuum Publishers, London, pp. 282–300.

Carlsen, J. and Butler, R. (2011) *Island Tourism: Sustainable Perspectives.* CAB International, Wallingford, UK.

Cave, J. and Brown, K.G. (2012) Island Tourism: Destinations: an Editorial Introduction to the Special Issue. *International Journal of Culture Tourism and Hospitality Research* 6(2), 95–113.

Chen, C.M., Chen, S.H., Lee, H.T. and Tsai, T.H. (2016) Exploring destination resources and competitiveness – A comparative analysis of tourists' perceptions and satisfaction toward an island of Taiwan. *Ocean and Coastal Management* 119, 58–67.

Chen, R.J.C. (2006) Islands in Europe: development of an island tourism multi-dimensional model (ITMDM). *Sustainable Development* 14, 104–114.

Choi, H.S.C. and Sirakaya, E. (2005) Measuring residents' attitude toward sustainable tourism: Development of sustainable tourism attitude scale. *Journal of Travel Research* 43(4), 380–394.

Choy, D.J. (1992) Life cycle models for Pacific island destinations. *Journal of Travel Research* 30(3), 26–31.

Conlin, M. and Baum, T. (eds) (1995) *Island Tourism: Management Principles and Practice.* John Wiley & Son, New York.

Conlin, M.V. and Baum, T. (1996) Macro aspects of productivity planning for the hospitality industry. In: Johns, N. (ed.) *Productivity Management in Hospitality and Tourism.* Cassell, London, pp. 55–66.

Craigwell, R. (2007) Tourism competitiveness in Small Island developing states (No. 2007/19). Research paper, UNU-WIDER, United Nations University (UNU).

Croes, R.R. (2006) A paradigm shift to a new strategy for small island economies: Embracing demand side economics for value enhancement and long term economic stability. *Tourism Management* 27(3), 453–465.

Croes, R. (2011) Measuring and Explaining Competitiveness in the Context of Small Island Destinations. *Journal of Travel Research* 50(4), 431–442.

Croes, R. (2013) Tourism specialization and economic output in small islands. *Tourism Review* 68(4), 34–48.

Croes, R. (2016) Connecting Tourism Development with Small Island destinations and with the well-being of the Island Residents. *Journal of Destination Marketing and Management* 5, 1–4.

Croes, R. and Rivera, M. (2016) *Poverty Alleviation through Tourism Development – A Comprehensive and Integrated Approach.* CRC Press, Boca Raton, Florida.

Croes, R. and Semrad, K.J. (2013) The relevance of cultural tourism as the next frontier for small island destinations. *Journal of Hospitality and Tourism Research*, DOI: 1096348013491599.

Crouch, G.I. (2010) Destination competitiveness: An analysis of determinant attributes. *Journal of Travel Research* DOI; 10.1177/0047287510362776.

Crouch, G.I. and Ritchie, B.J.R. (2012) Destination competitiveness and its implications for host-community QOL. In: Uysal, M., Perdue, R. and Sirgy, M.J. (eds) *Handbook of Tourism and Quality-of-life Research: Enhancing the lives of tourists and residents of host communities.* Springer, Dordrecht, the Netherlands, pp. 491–514.

Cuccia, T. and Rizzo, I. (2011) Tourism seasonality in cultural destinations: Empirical evidence from Sicily. *Tourism Management* 32(3), 589–595.

Daye, M., Chambers, D. and Roberts, S. (2008) *New Perspectives in Caribbean Tourism.* Routledge, London.

Dodds, R. and Joppe, M. (2009) The demand for, and participation in corporate social responsibility and sustainable tourism –implications for the Caribbean. *ARA Journal of Travel Research* 2(1), 1–24.

Dupeyras, A. and MacCallum, N. (2013) Indicators for measuring competitiveness in tourism: A guidance document. OECD Tourism Papers, 2013/02, OECD Publishing.

Duval, D.T. (2004) *Tourism in the Caribbean: Trends, Development and Prospects.* Routledge, London.

Dwyer, L. and Kim, C. (2003) Destination competitiveness: determinants and indicators. *Current Issues in Tourism* 6(5), 369–414.

European Commission (2016) *The European Tourism Indicator System.* Publications Office of the European Union, Luxembourg.

Farsari, Y., Butler, R. and Prastacos, P. (2007) Sustainable tourism policy for Mediterranean destinations: issues and interrelationships. *International Journal of Tourism Policy* 1(1), 58–78.

Gayle, D.J. (2002) *Island Tourism and Sustainable Development: Caribbean, Pacific, and Mediterranean experiences.* Greenwood Publishing Group, Westport, Connecticut.

Hampton, M.P. and Christensen, J. (2007) Competing industries in islands: a new tourism approach. *Annals of Tourism Research* 34(4), 998–1020.

Hampton, M.P. and Jeyacheya, J. (2013) *Tourism and Inclusive Growth in Small Island Developing States.* Commonwealth Secretariat, London.

Hamzah, A. and Hampton, M.P. (2013) Resilience and non-linear change in island tourism. *Tourism Geographies* 15(1), 43–67.

Harrison, D. (2003) *Pacific Island Tourism.* Cognizant Communication Corporation, New York.

Ioannides, D. and Holcomb, B. (2003) Misguided policy initiatives in small-island destinations: why do up-market tourism policies fail? *Tourism Geographies* 5(1), 39–48.

Ioannides, D., Apostolopoulos, Y. and Sonmez, S. (2001) *Mediterranean Islands and Sustainable Tourism Development. Practices, Management and Polices.* Continuum, London.

Jayawardena, C. (2005) *Caribbean Tourism: Visions, Missions and Challenges.* Ian Randle Publishers, Kingston, Miami.

Jensen, O. (2014) Approaches for the evaluation of visitor experiences at tourist attractions. In: Prebensen, N., Chen, J. and Uysal, M. (eds) *Creating Experience Value in Tourism.* CAB International, Wallingford, UK, pp. 139–156.

Kenchington, R.A. (1989) Tourism in the Galapagos Islands: the dilemma of conservation. *Environmental Conservation* 16(3), 227–232.

Kerstetter, D. and Bricker, K. (2012) Relationship between carrying capacity of small island tourism destinations and quality-of-life. In: Uysal, M., Perdue, R. and Sirgy, J. (eds) *Handbook of Tourism and Quality-of-Life: The Missing Links.* Springer, the Netherlands, pp. 445–462.

Kim, Y. and Uysal, M. (2002) Challenge for the Future: Tourism Strategies for Small-Islands. In: Apostolopoulos, Y. and Gayle, D.J. (eds) *Island Tourism and Sustainable Development.* Praeger, Westport, Connecticut, pp. 273–292.

Kokkranikal, J., McLellan, R. and Baum, T. (2003) Island tourism and sustainability: A case study of the Lakshadweep Islands. *Journal of Sustainable Tourism* 11(5), 426–447.

Lewis-Cameron, A. and Roberts, S. (2010) *Marketing Island Destinations.* Routledge, London.

Lim, C.C. and Cooper, C. (2009) Beyond sustainability: optimizing island tourism development. *International Journal of Tourism Research* 11(1), 89–103.

MacLeod, D.V.L. (2004) *Tourism, Globalization and Cultural Change: An Island Community Perspective.* Channel View Publications, Clevedon, UK.

Manning, R., Wang, B., Valliere, W., Lawson, S. and Newman, P. (2002) Research to estimate and manage carrying capacity of a tourist attraction: a study of Alcatraz Island. *Journal of Sustainable Tourism* 10(5), 388–404.

McElroy, J.L. (2006) Small island tourist economies across the life cycle. *Asia Pacific Viewpoint* 47(1), 61–77.

Meyer-Arendt, K.J. and Lew, A.A. (2013) New perspectives on tropical coastal and island tourism development. *Tourism Geographies* 15(1), 1–2.

Michael Hall, C. (2010) Island destinations: a natural laboratory for tourism: introduction. *Asia Pacific Journal of Tourism Research* 15(3), 245–249.

Modica, P. (2015) *Sustainable Tourism Management and Monitoring. Destination, Business and Stakeholder Perspectives*. FrancoAngeli, Milan.

Reddy, M.V. (2008) Sustainable tourism rapid indicators for less-developed islands: an economic perspective. *International Journal of Tourism Research* 10(6), 557–576.

Ridderstaat, J., Oduber, M., Croes, R., Nijkamp, P. and Martens, P. (2014) Impacts of seasonal patterns of climate on recurrent fluctuations in tourism demand: Evidence from Aruba. *Tourism Management* 41, 245–256.

Rodríguez, J.R.O., Parra-López, E. and Yanes-Estévez, V. (2008) The sustainability of island destinations: Tourism area life cycle and teleological perspectives. The case of Tenerife. *Tourism Management* 29(1), 53–65.

Sahli, M., Dwyer, L., Maupertuis, M.A. and Nowak, J.J. (2007) Special Issue: Island tourism economies and sustainable development. *Tourism Economics* 13(4), 494–684.

Sarrasin, B. (2013) Ecotourism, poverty and resources management in Ranomafana, Madagascar. *Tourism Geographies* 15(1), 3–24.

Scheyvens, R. (2011) *Tourism Poverty*. Routledge, London.

Scheyvens, R. and Momsen, J. (2008) Tourism in small island states: From vulnerability to strengths. *Journal of Sustainable Tourism* 16(5), 491–510.

Sharpley, R. (2001) Sustainability and the political economy of tourism in Cyprus. *Tourism (Zagreb)* 49(3), 241–254.

Sheldon, P.J. (2005) The Challenges to sustainability in island tourism. *Occasional Paper*, 2005, 01. School of Travel Industry Management, University of Hawaii, Mano.

Stratigea, A. and Katsoni, V. (2015) A strategic policy scenario analysis framework for the sustainable tourist development of peripheral small island areas – the case of Lefkada-Greece Island. *European Journal of Futures Research* 3(1), 1–17.

Su, M.M., Wall, G. and Jin, M. (2016) Island livelihoods: Tourism and fishing at Long Islands, Shandong Province, China. *Ocean and Coastal Management* 122, 20–29.

Tanguay, G.A., Rajaonson, J. and Therrien, M.C. (2013) Sustainable tourism indicators: Selection criteria for policy implementation and scientific recognition. *Journal of Sustainable Tourism* 21(6), 862–879.

Tolkach, D. and King, B. (2015) Strengthening Community-Based Tourism in a new resource-based island nation: Why and how? *Tourism Management* 48, 386–398.

Tsartas, P. (2003) Tourism development in Greek insular and coastal areas: sociocultural changes and crucial policy issues. *Journal of Sustainable Tourism* 11(2–3), 116–132.

UN (2014) *Trends in Sustainable Development: Small Island Developing States (SIDS)*. United Nations, New York, 37 pp.

UNWTO (2004) *Guide on Indicators*. WTO, Madrid, Spain.

Uysal, M., Woo, E. and Signal, M. (2012a) The tourist area life cycle (TALC) and its effect on the quality of life (QOL) of destination community. In: Uysal, M., Perdue, R. and Sirgy, J. (eds) *Handbook of Tourism and Quality-of-Life: The Missing Links*. Springer, Dordrecht, the Netherlands, pp. 423–444.

Uysal, M., Perdue, R. and Sirgy, J. (eds) (2012b) *Handbook of Tourism and Quality-Of-Life Research: Enhancing the Lives of Tourists and Residents of Host Communities*. Springer, Dordrecht, the Netherlands, 702 pp.

Veríssimo, D., Fraser, I., Groombridge, J., Bristol, R. and MacMillan, D.C. (2009) Birds as tourism flagship species: a case study of tropical islands. *Animal Conservation* 12(6), 549–558.

WEF (2015) *The Travel and Tourism Competitiveness Report 2015*. World Economic Forum, Geneva, Switzerland.

Index

Note: Page numbers in **bold** type refer to **figures**; Page numbers in *italic* type refer to *tables*; Page numbers followed by 'n' refer to notes